Electricity 1

DEVICES
CIRCUITS
MATERIALS

THIRD EDITION

Electricity 1

DEVICES
CIRCUITS
MATERIALS
THIRD EDITION

THOMAS S. KUBALA

COPYRIGHT © 1981
BY DELMAR PUBLISHERS INC.

10 9 8 7 6 5 4 3 2

ISBN: 0-8273-1357-8
LIBRARY OF CONGRESS CATALOG CARD NUMBER: 79-93322

Printed in the United States of America
Published Simultaneously in Canada by
Delmar Publishers, A division of
Van Nostrand Reinhold, Ltd.

DELMAR PUBLISHERS INC. • ALBANY, NEW YORK 12205

Preface

Electrical energy is the most important and widely used form of energy today. The electrician provides the specialized expertise and techniques required to put electricity to work for the benefit of all of us. The electrician must have a thorough knowledge of electrical devices and materials, and must understand the theory and characteristics of electrical circuits. The electrician is expected to apply this knowledge to practical circuit installations such as those found in homes, businesses and industries.

ELECTRICITY 1 provides the student with the necessary basic information on electrical devices, circuits, and materials. In addition, this text serves as the basis for further study. The student should realize that the study of electricity is a continuing process. The electrical industry is constantly introducing new and improved devices and materials, while electrical codes undergo periodic revisions to upgrade safety and quality in electrical installation. While studying the units in this text, the student will find it valuable to refer to the most recent edition of the National Electrical Code, published by the National Fire Protection Association in the interest of life and property protection. State and local regulations, as they apply to a specific area, should also be consulted when making an actual installation.

The emphasis in content of ELECTRICITY 1 is on the understanding of electrical concepts and the use of devices and materials. The student is advised that electron movement (from negative to positive) is used in this text to define current direction.

The text is easy to read and comprehend, and the topics are presented in a logical sequence. The mathematical level involves only simple algebra. The student should not find it difficult to understand and calculate the simple mathematical equations required in this study.

The first nine units of ELECTRICITY 1 present basic electrical theory, fundamental circuits, and information on energy and power, batteries, and electrical conductors. Units 11–14 include magnetic theory and coverage of electromagnetism, generators, and motors. In Units 16–19, practical circuits are presented to assist the student in learning to apply the theory to common situations. Units 10, 15, and 20 are Summary Review units in which a comprehensive examination of previous units is given. The purpose of these review units is to evaluate the knowledge and understanding of the theory acquired by the student to that point. The Achievement Review at the end of each of the other units requires thought and comprehension on the part of the student. These reviews are essential to the learning process required by this text.

All students of electricity will find this text useful, especially those in electrical apprenticeship programs, trade and technical schools, and various occupational programs. Students should also find it beneficial to use this text in conjunction with the DIRECT CURRENT FUNDAMENTALS text, also published by Delmar in its electrical trades series.

An Instructor's Guide is available. An outline of each unit is provided, including suggested class demonstrations, to assist the instructor in planning the presentation to the students. In addition, answers are provided for each Achievement Review and the three Summary Reviews. Complete solutions for the more complex problems are also included.

Dr. Thomas S. Kubala, the author of ELECTRICITY 1, is the President of Thomas Nelson Community College, Hampton, Virginia. Prior to accepting this post, he was Dean of the College, Anne Arundel Community College, Arnold, Maryland, having previously held such positions there as Dean of Career Programs, Associate Professor of Engineering and Director of Electrical Technology.

Dr. Kubala received an AAS degree in Electrical Technology from Broome Technical Community College, Binghamton, New York. He earned a BS degree in Electrical Engineering from the Rochester Institute of Technology, Rochester, New York, and an MS degree in Vocational-Technical Education from the State University of New York at Oswego, Oswego, New York. He earned his doctoral degree from the University of Maryland, College Park, Maryland.

In addition to his extensive background in technological education, Dr. Kubala has also had several years of experience in industry with responsibilities in the fields of aerodynamics, and electrical drafting, design, testing and evaluation. He has been a member of several professional organizations during his career.

Dr. Kubala is also the author of ELECTRICITY 2, PRACTICAL PROBLEMS IN MATHEMATICS FOR ELECTRICIANS, and CIRCUIT CONCEPTS: DIRECT AND ALTERNATING CURRENT, all published by Delmar.

A current catalog including prices of all Delmar educational publications is available upon request. Please write to:

Catalog Department
Delmar Publishers Inc.
50 Wolf Road
Albany, New York 12205

ELECTRICAL TRADES

The Delmar series of instructional material for electrical trades includes the texts, text-workbooks, and related information workbooks listed below. Each text features basic theory with practical applications and student involvement in hands-on activities.

ELECTRICITY 1

ELECTRICITY 2

ELECTRICITY 3

ELECTRICITY 4

ELECTRIC MOTOR CONTROL

ELECTRIC MOTOR CONTROL LABORATORY MANUAL

ALTERNATING CURRENT FUNDAMENTALS

DIRECT CURRENT FUNDAMENTALS

ELECTRICAL WIRING – RESIDENTIAL

ELECTRICAL WIRING – COMMERCIAL

ELECTRICAL WIRING – INDUSTRIAL

PRACTICAL PROBLEMS IN MATHEMATICS FOR ELECTRICIANS

EQUATIONS BASED ON OHM'S LAW

P = Power in Watts

I = Intensity of Current in Amperes

R = Resistance in Ohms

E = Electromotive Force in Volts

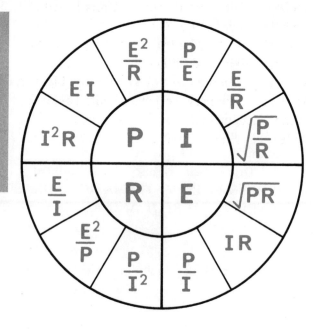

Contents

1 Introduction

OBJECTIVES

After studying this unit, the student will be able to

- list the areas of work in which the apprentice electrician becomes involved.
- discuss the ethics of the electrical trade.
- describe the apprentice program as it applies locally, and discuss the values of any apprentice program.

When beginning a new program of study, an individual should be thoroughly familiar with the nature of the program, and its values and requirements. This is especially important when the program involves training for a lifelong occupation.

DESCRIPTION OF THE TRADE

The electrical trade is one of the basic trades in the construction industry. It is a trade in which individual ability and skill is recognized and rewarded. The trade involves the following areas: electrical installation in new buildings, rewiring of old buildings, electrical maintenance and repair, and troubleshooting of electrical equipment and installations. Many of these areas are also basic to the power and electronics fields.

The work involved in all fields often is so closely related to the technical and theoretical concepts of electricity that only a trained person can do the job. This is especially true in the field of electronics. Since more and more electronic equipment is being used, it is expected that the electrician can install and maintain this equipment. Therefore, it is necessary for the electrical apprentice to acquire the related technical information.

WORKING CONDITIONS IN THE TRADE

The surroundings and working conditions of the electrical trade are favorable to the worker. The trade offers opportunities for indoor and outdoor work. Working hours and conditions of the trade permit the electrical worker to find pleasure in doing a first-class job. The journeymen on many jobs have the opportunity to deal with customers; therefore, personal conduct of the experienced worker affects future advancement of the trade and industry. The electrical trade requires a high degree of responsibility on the part of the trained technician since this person has the responsibility for the interconnection and construction of complex electrical systems. These systems are controlled by state and local building codes, and the National Electrical Code. As a result, the work requires skilled technicians.

OPPORTUNITIES IN THE TRADE

The general public's interest in building construction at the present time demands a greater number of more highly trained electricians. The modern home, office, and factory require a higher degree of proficiency in electrical work. The constant increase in new types of construction, new electrical equipment, and new uses for electrical equipment offers increasing employment opportunities for qualified electricians. The ever-increasing use of electronic equipment in the power field has shown the need for advanced training of electricians.

Technological advances have created new improvements, new ideas, and new processes. It is necessary for the apprentice to be familiar with these developments in order to advance in the electrical profession. The increased use of this information by the electrician makes the electrical trade more interesting and desirable. The opportunity is open for the apprentice to become a first-class journeyman by understanding new phases of the electrical field. A first-class journeyman can advance to the position of foreman or contractor. The electrical trade is in need of individuals with a complete knowledge of the practical and technical phases of the trade including those who can supervise workers on the job.

Some of the fields which offer opportunities are: electrical construction, line construction, cable installation, signaling systems, light and power systems, electrical motor maintenance and repair, equipment and appliance servicing, and industrial electronics. Due to the increased needs of our society, new opportunities are developing very rapidly.

ETHICS OF THE TRADE

Electricians are judged by the quality of their work and by their attitude toward fellow workers, employers and the public. A good electrician takes pride in doing high-quality work and gives an honest day's work for an honest day's pay. An accurate and complete job is expected in every activity.

QUALIFICATIONS FOR EMPLOYMENT

Educational

The apprentice should be a high school graduate or equivalent. A trainee should be eager to learn the skills and technical information necessary for success in the electrical trade. The apprentice is expected to have a working knowledge of mathematics since this aids in the understanding of the important and necessary electrical formulas.

Physical

The apprentice must meet the requirements of a physical examination. A person must be strong enough to perform certain duties since the trade requires a considerable amount of moving about, climbing, and working under conditions which require muscular action. The general health of the apprentice should be good.

General

The apprentice must like to work with electrical equipment and should be interested in the general theory of electricity. The trainee must like to work with others in a cooperative manner. Often, it is necessary for electricians to work in pairs and also with individuals in other trades. Electrical apprenticeship requires a liking for indoor as well as outdoor work, and a willingness to do a fair share of manual labor.

VALUES OF THE APPRENTICE PROGRAM

- Apprenticeship is an educational experience for the apprentice.
- The apprentice training program provides for training on an organized basis.
- A controlled apprenticeship brings together the fundamental factors which are necessary to produce a skilled technician.
- Apprenticeship is the most practical and efficient means of training a skilled technician.
- An apprenticeship program is of benefit to the trainee, the employer, the union, and society because all benefit from better workmanship.
- The successful electrician profits according to his or her knowledge and skill. It is an advantage to have the highest qualities possible.

THE RESPONSIBILITIES OF THE APPRENTICE

Educational programs and the work experience provide the apprentice with the opportunity to acquire the knowledge and skill necessary to become a skilled technician. It is the trainee's responsibility to make the most of these opportunities.

The apprentice is expected to take an interest in his or her work, to have a desire to learn, to fit into the employer's organization, to know the correct relationship of the apprentice to the journeyman, to plan and organize his or her work efficiently, to be resourceful, and to know how to conserve materials.

The trainee is further expected to be punctual, to maintain good health, to develop initiative and leadership, to cooperate in every way, to be neat in personal appearance, and to practice safe working procedures at all times.

The apprentice is expected to keep informed regarding new facts, new ideas, and new procedures of the trade. Because an apprentice is also expected to continue learning while earning, the trainee must be prepared to attend school to obtain the necessary technical and related instruction.

THE PROGRAM OF RELATED INSTRUCTION

Generally, the apprenticeship agreement requires the apprentice to attend classes in related subjects for a minimum number of hours. The length of the apprenticeship period in the electrical trade is normally five years. In certain localities, time spent in related instruction is not classified as work time and is not paid for, while in other localities, school attendance is considered work time so that the apprentice receives pay at the prevailing wage rate.

The program of instruction consists of courses based on divisions of work within the trade such as residential wiring, commercial wiring, industrial plant wiring, and maintenance and repair. Each course includes such information as trade science, trade mathematics, and trade theory and practice.

If the apprentice enters a related instruction program at the time the course is being taught, he or she will obtain instruction in the normal manner by attending classes. If the related instruction course is not being given at the time the apprentice enters the program, this information must be acquired through self-study, under the supervision and with the assistance of the instructor. Apprentices are expected to provide their own necessary materials, such as textbooks, notebooks and workbooks, as advised by the instructor.

Electron Theory and Ohm's Law

2

OBJECTIVES

After studying this unit, the student will be able to

- list the fundamental properties of matter.
- describe the structure of an atom.
- explain the basic electrical concepts of current, voltage, resistance, and electrical polarity.
- define Ohm's Law.

MATTER

Anything which occupies space and has weight is called *matter*. All liquids, gases, and solids are examples of matter in different forms. Matter itself is made up of smaller units called atoms.

ATOMS

An *atom* resembles the solar system with the sun as the center around which a series of planets revolve, figure 2-1. In the atom there is a relatively large mass at the center called the *nucleus*. Around the nucleus are electrons which revolve in orbital patterns.

Fig. 2-1 Atomic structure.

ELECTRICAL CHARGE

A material is said to have an *electrical charge* when it attracts or repels another charged material. A material may have either a positive or a negative electrical charge.

Two objects with positive charges repel each other. Two objects with negative charges also repel each other. Two objects with unlike charges attract each other.

PROTONS AND NEUTRONS

Part of the nucleus of an atom is made up of protons. Each *proton* has a positive electrical charge and attracts electrons; neutrons form the remainder of the nucleus. *Neutrons* are electrically neutral. They can neither attract nor repel other electrical charges.

ELECTRON

One or more electrons revolve continuously about the nucleus of an atom (just as the planets revolve about the sun). *Electrons* possess a negative electrical charge and are very much lighter in weight than protons. All electrons are alike regardless of the atoms of which they are a part. An atom contains the same number of electrons as there are protons. For example, the aluminum atom has 13 electrons and 13 protons.

CURRENT

Electrons in motion result in an electrical current. Copper wire often is used to carry electrical current (moving electrons). For each atom of copper in the wire, electrons are revolving about the nucleus. When electrical pressure (voltage) from a battery or generator is applied, it is possible to force these electrons out of their circular paths and cause them to pass from atom to atom along the length of the wire (conductor).

The greater the number of electrons passing a given point in a circuit the greater the intensity of the current. The intensity of an electrical current is measured in *amperes* (A). The instrument used to measure current is called an *ammeter*. The letter (I) is used to represent the amount of current in a circuit.

Current Types

Direct current (dc) is the movement of electrons in one direction in a conductor.

Pulsating direct current is a current in one direction which varies in intensity at a regular interval of time.

Alternating current (ac) is a current which changes in direction and intensity at a regular interval of time.

The three types of current are shown in figure 2-2.

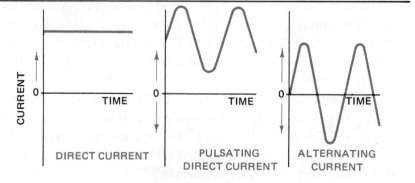

Fig. 2-2 Types of electrical current.

VOLTAGE

A closed circuit and a source of electrical pressure are necessary to produce an electrical current. Electrical pressure, known as *voltage,* or *potential,* is obtained from many sources. Generators are widely used for high-powered ac and dc installations. Storage batteries are used extensively for dc power in automobiles and aircraft. Photo-electric cells convert light energy into electrical energy. These cells are used as voltage sources in light-operated devices. A *thermocouple,* which consists of a junction of two unlike metals, generates a low voltage when heated. Of all the voltage sources mentioned, the generator is most important because of the magnitude of its commercial applications.

The letter (E) is used to represent a voltage. The *volt* (V) is the unit used to express the quantity of electric potential. The instrument used to measure voltage is the *voltmeter.*

ELECTRICAL POLARITY

All dc sources of electrical potential have two terminals to which electrical devices are connected. These terminals have what is known as *electrical polarity.* One terminal is the positive terminal, while the other is called the negative terminal. Electrons flow through the device from the negative terminal of the source to the positive terminal of the source. The source maintains a supply of electrons on its negative terminal.

RESISTANCE

The property of a material which causes it to oppose the movement of electrons is called *resistance.* All materials have some resistance. Materials which offer little resistance to electron movement are called *conductors.* Those which offer high resistance are called *nonconductors* or *insulators.*

Resistance is measured in *ohms.* The symbol for ohms is the Greek letter omega (Ω). This symbol, representing ohms, and the letter (R), representing resistance, are used in formulas. The instrument used to measure resistance is called an *ohmmeter.*

OHM'S LAW

It is extremely important to understand the methods used to control the amount of current in a circuit. A simple formula, Ohm's Law, is used to show the relationship of current, voltage, and resistance. *Ohm's Law* states that in any electrical circuit the current is directly proportional to the voltage applied to the circuit and is inversely proportional to the resistance in the circuit. Note that both resistance and voltage affect the current.

According to Ohm's Law, when the resistance of a circuit is constant, the current can be changed by changing the voltage: current will increase when the voltage is increased, and current will decrease when the voltage is decreased. Similarly, when the voltage is constant, current will increase when the resistance is decreased, and current will decrease when resistance is increased.

The exact relationship of voltage, current, and resistance is expressed by the equation for Ohm's Law:

$$I = \frac{E}{R}$$

Where **I** = intensity of current in amperes
E = quantity of electrical potential in volts
R = amount of resistance in ohms

Two other forms of Ohm's Law are:

$$E = IR \text{ and } R = \frac{E}{I}$$

ACHIEVEMENT REVIEW

1. Explain the meaning of voltage, current, and resistance.

    _____

    _____

    _____

2. State Ohm's Law and write three forms of Ohm's Law using equations.

    _____

    _____

    _____

    _____

3. What instruments are used to measure voltage, current, and resistance?

    _____

    _____

    _____

4. What units of measure are used for voltage, current, and resistance?

    _____

    _____

    _____

5. A trouble light has a resistance of 12 ohms and is rated at 1/2 ampere. What voltage must be applied to obtain the rated current?

    _____

6. What current is taken by a heater with a resistance of 24 ohms when connected to a 120-volt supply? _____

_____

7. Determine the resistance of a lamp that draws 3 amperes when connected to a 120-volt supply. _____

_____

8. If the lamp in problem 7 is connected to a 240-volt supply, what is the new value of current?   (Assume there is no change in resistance as the temperature of the lamp changes.) _____

_____

9. An 8-ohm resistor is connected to a 120-volt circuit. What current will it draw?

_____

_____

10. If 60 volts are applied to an 8-ohm resistor, what is the value of current through the resistor? _____

_____

3 Series Circuits

OBJECTIVES

After studying this unit, the student will be able to

- describe the basic relationships of voltage, current, and resistance in a series circuit.
- apply Ohm's Law to determine unknown quantities.

A knowledge of certain basic rules in the operation of series, parallel, and series-parallel circuits is of importance in developing a facility for locating faults in electrical equipment. A grasp of electrical problems is, in fact, impossible without this knowledge.

A *series circuit* is one in which devices are connected so that there is only one path for current. The direction of the current in the wire is the same as the direction of electron movement. Figure 3-1 illustrates three lamps connected in series with a voltage source.

Fig. 3-1 **Three lamps connected in series.**

VOLTAGE

The total voltage applied to a series circuit is distributed across the various devices of the circuit in a series of *voltage drops*.

The three equal resistors shown in figure 3-2 are connected in series. The voltage across each device is equal to one-third of the total voltage. In figure 3-3, page 12, the voltage across each resistor is in proportion to the resistance of the device.

It can be seen from the figures that the sum of the voltages across the individual devices is equal to the total applied voltage. This leads to the following important rule for a series circuit:

Fig. 3-2 **Voltage and current distribution: resistors of equal value in series.**

Fig. 3-3   Voltage and current distribution: resistors of unequal values in series.

The sum of the voltage drops across individual resistors in a series circuit is equal to the total applied voltage. In other words,

$$E_T = E_1 + E_2 + E_3.$$

CURRENT

Since only one path for current exists, the current through all devices in the circuit is the same. This statement can be expressed as,

$$I_T = I_1 = I_2 = I_3$$

Where I_T = total current

I_1 = current through device 1

I_2 = current through device 2

I_3 = current through device 3

RESISTANCE

The total resistance of a series circuit is equal to the sum of the resistances of all parts of the circuit. The total resistance in figure 3-1 is the resistance from terminal A to terminal B with the voltage source disconnected.

In equation form,

$$R_T = R_1 + R_2 + R_3$$

Where R_T = total circuit resistance

R_1 = resistance of device 1

R_2 = resistance of device 2

R_3 = resistance of device 3

An alternate path of very low resistance in a circuit is called a *short circuit*. For example, if the two wires leading to a lamp should come in contact with each other, a path of practically zero resistance is formed. When this happens, there is a very large current in the wires leading to the place of contact, and overheating of the wires will occur.

An *open circuit* occurs when some part of a circuit is either open, such as a switch, or malfunctioning, such as a burned-out fuse or a broken wire. There is no current anywhere in the circuit. However, the source voltage must be accounted for. If a voltmeter is used at an open point in a circuit, it will indicate the source voltage.

ACHIEVEMENT REVIEW

1. Four devices are connected in series across 110 volts dc. The devices fail to operate. A voltmeter connected in succession across each device reads zero across the first three devices and 110 volts across the fourth device. What circuit fault is indicated?

   _____

2. Four devices are connected in series across 120 volts. One device fails to operate. The voltage across each of the other devices is 40 volts. What circuit fault is indicated?

   _____

3. State three characteristics of a series circuit. _____

   _____

   _____

   _____

4. Find the voltage drop across a 10-ohm resistor, if the current through the resistor is 1.7 amperes. _____

   _____

5. Find the resistance of a resistor if the voltage drop across it is 51 volts, and the current through it is 3 amperes. _____

   _____

6. Solve for the unknown

   I_T =

   R_T =

$R_1 = 5 \, \Omega$	$R_2 = 10 \, \Omega$	$R_3 = 15 \, \Omega$
$I_1 =$ ___	$I_2 =$ ___	$I_3 =$ ___
$E_1 =$ ___	$E_2 =$ ___	$E_3 =$ ___

   150 V

7. Solve for the unknown values.

   E_G =

   I_T = 10 amperes

   R_T =

$E_1 =$ ___	$E_2 = 80$ volts
$R_1 = 4 \, \Omega$	$R_2 =$ ___
$I_1 =$ ___	$I_2 =$ ___

   E_G

4 Parallel Circuits

OBJECTIVES

After studying this unit, the student will be able to

- describe the characteristics of parallel circuits.
- demonstrate a procedure for solving parallel circuit problems.

Because of their unique characteristics, parallel circuits are more widely used than any other type of circuit. The distribution of power in a large city is accomplished by a maze of feeder lines all connected in parallel.

VOLTAGE

The circuit shown in figure 4-1 is an example of a simple parallel circuit. Note that each resistor is placed directly across the main source of voltage. This causes each device to operate at the same voltage as the source. A device should never be placed in a parallel circuit if it has a voltage rating less than the source voltage.

Fig. 4-1 Unequal resistors connected in parallel.

The fact that all devices in a parallel circuit operate at the same voltage is expressed by the following equation:

$$E_T = E_1 = E_2 = E_3$$

Where E_T = total voltage

E_1 = voltage across device 1

E_2 = voltage across device 2

E_3 = voltage across device 3

CURRENT

The devices in a parallel circuit operate independently of one another. Each device takes current in accordance with its resistance. The number of separate paths for current is equal to the number of devices in parallel. The total current in a parallel circuit is equal to the sum of the currents in the separate devices. The equation that expresses this statement is:

$$I_T = I_1 + I_2 + I_3$$

Where I_T = total current

I_1 = current through device 1

I_2 = current through device 2

I_3 = current through device 3

RESISTANCE

It is apparent, from studying the current equation, that by adding more parallel branches to the circuit the total current will increase. Since the total increases while the source voltage remains constant, Ohm's Law shows that the total circuit resistance decreases.   Therefore, an *increase* in parallel branches results in a *decrease* in total resistance.

Equal Resistors

In a parallel circuit which consists of devices with equal resistance, figure 4-2, the total circuit resistance is numerically equal to the resistance value of one device divided by the number of devices connected in parallel. Expressed as an equation, this statement becomes,

Fig. 4-2 Equal resistors connected in parallel.

$$R_T = \frac{R}{N} = \frac{15}{3} = 5 \text{ ohms}$$

Where R_T = total resistance in ohms

R = resistance of one of the equal valued resistors in ohms

N = number of parallel resistors

Unequal Resistance

In practice, parallel circuits with resistors having unequal values are more frequently used than parallel circuits with resistors having equal values. No simple rule applies in this case because each resistor takes a different value of current for the same applied voltage.

To find the total resistance of a parallel circuit, apply a known source voltage to the circuit and then determine the total current. Ohm's Law is then used to find the total resistance.

$$R_T = \frac{E_T}{I_T}$$

Where E_T = applied voltage in volts

I_T = total current in amperes

R_T = total circuit resistance in ohms

The total circuit resistance also can be found by the use of the following formula. This formula may be applied to any parallel circuit with any number of parallel branches. Known as the "reciprocal" formula, it is expressed as,

$$\frac{1}{R_T} = \frac{1}{R_1} + \frac{1}{R_2} + \frac{1}{R_3}$$

Where R_T = total resistance

R_1 = resistance of device 1

R_2 = resistance of device 2

R_3 = resistance of device 3

For example, find the total resistance of the circuit in figure 4-1.

$$\frac{1}{R_T} = \frac{1}{3} + \frac{1}{6} + \frac{1}{8}$$

Lowest common denominator is 24

$$\frac{1}{R_T} = \frac{8}{24} + \frac{4}{24} + \frac{3}{24}$$

$$\frac{1}{R_T} = \frac{8+4+3}{24} = \frac{15}{24}$$

Solving for R_T

$$15R_T = 24$$

$$R_T = \frac{24}{15} = \frac{8}{5} = 1.6 \text{ ohms}$$

A simple method of solving circuits consisting of only *two* resistors in parallel (with either equal or unequal values) is called the "product over the sum" method.

Example: A 3-ohm resistor and a 6-ohm resistor are connected in parallel. Determine their combined resistance.

$$R_T = \frac{R_1 \times R_2}{R_1 + R_2} = \frac{3 \times 6}{3 + 6} = \frac{18}{9} = 2 \text{ ohms}$$

ACHIEVEMENT REVIEW

1. Four 12-ohm resistors are connected in parallel. Calculate the total circuit resistance.

   _____

2. Four resistors are connected in parallel. The resistance values are 4 ohms, 8 ohms, 12 ohms, and 16 ohms. Calculate the total circuit resistance.

   _____

   _____

3. The resistors mentioned in problem 2 are connected in parallel across a 120-volt dc supply.
   a. Calculate the current through each resistor.
   b. Find the total current.
   c. Find the total circuit resistance.

   _____

   _____

   _____

   _____

4. Determine the total resistance of a 5-ohm resistor and a 15-ohm resistor connected in parallel.

   _____

5. Find the value of R_2 if the total circuit resistance is 7.5 ohms.

   _____

   _____

   _____

   _____

   _____

6. The ammeters are indicating 4 amperes and 9 amperes as shown. Find the values of R_3 and R_T.

_____

_____

_____

_____

_____

5 Series-Parallel Circuits

OBJECTIVES

After studying this unit, the student will be able to

- explain the characteristics of series-parallel circuits.
- demonstrate a procedure for solving problems involving series-parallel circuits.

Often, it is necessary to combine series and parallel circuits to meet electrical requirements and to group devices in a load circuit to obtain a particular value of resistance. The grouping of devices in series-parallel circuits is also necessary in control circuits for auditorium and stage lighting as well as for motor control. In many instances, it is desirable to group voltage sources, particularly batteries, to obtain the correct voltage and current capacity.

The circuit shown in figure 5-1 is an example of a series-parallel circuit. In this circuit, lamps L_1 and L_2 constitute a parallel circuit. The rheostat R, used to control the current in this circuit, is in series with L_1 and L_2 as a group.

Figure 5-2 illustrates another series-parallel circuit. Resistors R_1 and R_2 are in parallel with respect to each other. Resistors R_3 and R_4 constitute another parallel combination. The parallel combination of R_1 and R_2 is in series with the parallel combination of R_3 and R_4.

In figure 5-3, the resistors are grouped in another type of series-parallel circuit. In this circuit, R_1 and R_3 are in series, and R_2 and R_4 are in series. The two series branches then are in parallel.

Fig. 5-1 A series-parallel circuit.

Fig. 5-2 A series-parallel circuit.

Fig. 5-3 A series-parallel circuit.

EQUIVALENT CIRCUITS

The methods which are used to determine current, voltage and resistance for series and parallel circuits apply to combination circuits as well. The solution of problems in series-parallel circuits is made easier by resolving these circuits into equivalent circuits.

Figure 5-4 is equivalent to figure 5-3. In this case, R_1 and R_3 are combined as a single resistance R_A, equal in value to the sum of R_1 and R_3. Similarly, R_B replaces R_2 and R_4. R_A and R_B then may be combined into one resistor, R_C, to result in the final equivalent circuit of figure 5-5. The total current in the original series-parallel circuit, figure 5-3, is equal to the current in the simple series circuit of figure 5-5.

Fig. 5-4 Equivalent circuit.   Fig. 5-5 Equivalent circuit.

CIRCUIT SOLUTION

Once the total resistance of a circuit is found, the total current, as well as the current in other parts of the circuit, may be determined according to Ohm's Law. In figure 5-6, the equivalent resistance of the parallel resistors R_2 and R_3 is 12 ohms. Therefore, figure 5-7 is the series circuit equivalent of figure 5-6, and the total resistance is:

$$R_T = R_1 + R_{2,3}$$
$$= 8\,\Omega + 12\,\Omega = 20\,\Omega.$$

The total current is: $I_T = \dfrac{120 \text{ volts}}{20 \text{ ohms}} = 6$ amperes.

The voltage across $R_{2,3}$ is: $I_T \times R_{2,3} = 6$ amperes $\times$ 12 ohms = 72 volts.

Fig. 5-6 A series-parallel circuit.   Fig. 5-7 Equivalent circuit.

Since $R_{2,3}$ is the equivalent resistance of the parallel combination of R_2 and R_3, the voltage across these resistors is 72 volts, as shown in figure 5-8.

Finally, the current through R_2 is:

$$I_2 = \frac{72 \text{ volts}}{20 \text{ ohms}} = 3.6 \text{ amperes,}$$

and the current through R_3 is:

$$I_3 = \frac{72 \text{ volts}}{30 \text{ ohms}} = 2.4 \text{ amperes.}$$

Fig. 5-8.

ACHIEVEMENT REVIEW

1. a. In figure 5-6, what circuit components are connected in series?

   _____

   b. What circuit components are in parallel with each other?

   _____

2. Assume that each resistor shown in figure 5-2 has a resistance of 100 ohms. Find the total circuit resistance. _____

   _____

   _____

   _____

3. a. In figure 5-9, what components are connected in series?

   _____

   b. What components are connected in parallel?

   _____

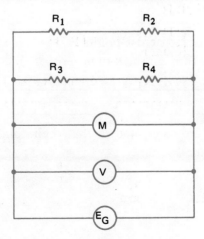

Fig. 5-9.

4. Determine the total current in the following circuit.

_____

_____

5. Find the current through the 6-ohm resistor for the circuit used in problem 4.

_____

_____

6. Determine the total resistance of the following circuit between points A and B.

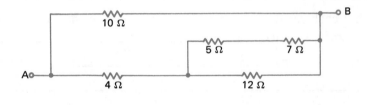

_____

_____

_____

_____

_____

_____

7. If 240 volts are connected across points A and B in the circuit shown in problem 6, what is the current through the 4-ohm resistor?

_____

_____

8. Five 4-ohm resistors are connected so that their combined resistance will equal 5 ohms. Draw the circuit diagram.

9. The two resistors in branch A-B are of equal value. What is the value of each resistor?

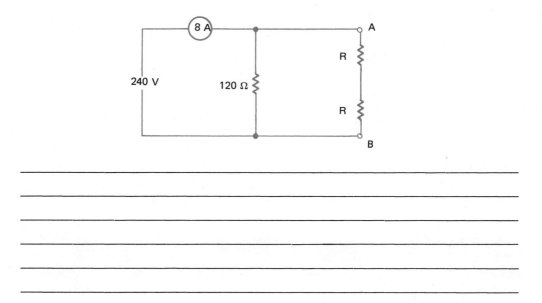

_____

_____

_____

_____

_____

_____

Electrical Energy and Power

6

OBJECTIVES

After studying this unit, the student will be able to

- discuss the relationship of work to power.
- apply the power and energy concepts to practical problems.

To insure proper operation, all electrical equipment is rated by the manufacturer. That is, the voltage and kind of current required are usually specified on the nameplate of the device. This information allows the consumer to compute the cost of operation before a purchase is made. A generator, for example, is rated for electrical power output. Damage to the generator results from operation at outputs in excess of this rating. An electrician cannot install an electric motor, and expect it to operate properly and safely, unless the horsepower requirements of the load are known. It is necessary to understand the exact meaning of all types of electrical ratings.

WORK

When an object is moved, there must be some force to make it move. When electrons flow in a circuit, there must be a force to make them flow. A *force* is that which produces or tends to produce motion or change in motion. *Energy* is the ability to do work. Therefore, when work is accomplished, energy is used or consumed.

If a weight is to be lifted, work is required. The unit of work is the *foot-pound* (ft·lb), which is the amount of work accomplished when a weight of one pound is lifted vertically one foot, or when a force of one pound acts through a distance of one foot. The amount of work done, measured in foot-pounds, is equal to the force in pounds multiplied by the distance in feet, or

$$\text{Work} = \text{Force} \times \text{Distance}$$

If a two-pound weight is lifted a distance of three feet, the work done is equal to 2 × 3, or 6 foot-pounds.

Work is not a function of time. An elevator motor does essentially the same amount of work in speeding a car to the top of a building as it does in having it rise slowly. Although the work is nearly the same, the motor must be much more powerful in the first instance than it is in the second.

POWER

Power is the rate of doing work. The faster a given amount of work is accomplished, the greater the power required. If the two-pound weight is raised three feet in one minute, more power is required than if the same weight were raised three feet in five minutes. Power is often expressed in foot-pounds per minute ($\frac{ft \cdot lb}{min}$).

$$\text{Power (foot-pounds per minute)} = \frac{\text{Work done (foot-pounds)}}{\text{Time (minutes)}}$$

A commonly used unit of power is the horsepower (hp).

$$\text{1 horsepower (hp)} = 33\,000 \frac{\text{foot-pounds}}{\text{minute}}$$

The *watt* (W) is used as the unit of power when dealing with electrical equipment. The instrument used to measure power is the *wattmeter*. When one ampere exists in a circuit due to a source of one volt, one watt of power is used in that circuit. In dc circuits the electrical power, in watts, may always be found by any of the following formulas in which I represents the number of amperes, R is the number of ohms, and E is the number of volts.

$$\text{Power} = I \times E \qquad \text{Power} = \frac{E^2}{R} \qquad \text{Power} = I^2 \times R$$

The *kilowatt* (kW) is a commonly used unit of electrical power. One thousand watts is equal to one kilowatt.

The energy consumed in electrical circuits is measured in *watthours* (Wh).

When one watt is used for one hour, the amount of energy consumed is one watt-hour. The *kilowatt-hour* (kWh) is equal to 1 000 watthours. In other words, a kilowatt-hour is the *energy* consumed when one kilowatt is used for one hour. When you pay an electric bill for your home, you are paying for the *energy* used, not power. The consumed energy is measured with an instrument called a *watthour meter*.

A simple formula can be used to determine the cost of the energy consumed.

$$\text{Cost} = \frac{\text{Watts} \times \text{Hours Used} \times \text{Cost per kWh}}{1\,000}$$

Determine the cost of operating a television set for 6 hours. The set is rated at 150 watts, and the cost of energy is at the rate of 3 cents per kWh.

$$\text{Cost} = \frac{150 \times 6 \times .03}{1\,000} = \$00.027 \text{ or } 2.7 \text{ cents}$$

Electrical power can be changed to mechanical power by an electric motor. If exactly as much power could be delivered by the motor as is supplied to it, then for each 746 watts of electrical power supplied to the motor one horsepower of mechanical power would be delivered.

$$746 \text{ watts} = 1 \text{ horsepower}$$

Actually, a motor is not 100 percent efficient. The power delivered is never equal to the power supplied. Some losses always occur due to internal motor resistance, bearing friction, and air friction. The power supplied to the motor must be greater than the power delivered to provide for these losses.

$$\text{Input} = \text{Output} + \text{Losses}$$

The percent efficiency of a machine is the ratio of the output power to the input power, and is always less than 100 percent.

$$\text{Percent Efficiency} = \frac{\text{Output power}}{\text{Input power}} \times 100$$

ACHIEVEMENT REVIEW

1. An electric soldering iron takes 5 amperes at 110 volts. What is the power used in watts? In kilowatts?

    _____

    _____

2. A device is rated at 1 100 watts. What current is required if it is operated at 110 volts? _____

3. A motor must lift an elevator car weighing 2 000 pounds to a height of 1 000 feet in four minutes, at a constant speed. What horsepower rating is required of the motor? _____

    _____

    _____

    _____

4. Find the cost of operating ten 100-watt lamps, at their rated voltage, for 11 hours at a cost of 10 cents per kilowatt-hour.

    _____

    _____

5. Determine the overall efficiency of a motor that delivers 2 hp to a load if it draws 7.5 amperes when connected to a 240-volt supply.

    _____

    _____

    _____

6. An electric toaster has a rating of 1 000 watts at 120 volts. What current will it draw? _____

7. The toaster in problem 6 (same heating element) is connected to a 240-volt circuit. What power will it use?

    _____

    _____

    _____

8. Determine the cost of operating a 2-watt electric clock for 365 days at 3 cents per kilowatt-hour. _____

    _____

9. A transformer primary circuit (input) draws 12 amperes when connected to a 2 400-volt source.   An ammeter connected in the secondary circuit (output) indicates 115 amperes at 240 volts.   Calculate the percent efficiency of the transformer.

_____

_____

_____

7 Batteries

OBJECTIVES

After studying this unit, the student will be able to

- describe the electrical characteristics of lead-acid storage cells.
- demonstrate how to test and charge storage batteries.
- list the most important aspects of storage battery maintenance.

Millions of batteries are used in America for automobiles, aircraft, portable lights, and emergency power installations. The ability to install, test, charge, and maintain storage batteries is an important asset to any well-qualified apprentice electrician.

CELLS

The basic unit of the battery is the *cell.* A battery is usually a group of separate cells connected in series. The number of cells used depends on the total voltage required.

Primary cells and secondary cells are two types of cells widely used in the electrical field. Primary cells are commonly known as dry cells. This type can be used only once. When discharged, they are commonly discarded. The secondary or storage-type cell, when discharged, can be recharged by passing direct current through it in the proper direction.

Two common types of storage cells are the nickel-cadmium cell and the lead-acid cell. The lead-acid cell is used extensively.

Lead-Acid Cells

The internal features of the lead-acid cell are shown in figure 7-1, page 32. Two groups of coated lead plates, known as *electrodes,* are immersed in a dilute solution of sulfuric acid known as the *electrolyte.* One group of plates forms the positive electrode, while the other forms the negative electrode. Glass, rubber, or other insulating materials are used as separators to keep these electrodes from making contact. Each cell container is provided with a vent and vent cap. These devices permit gases to leave the cell while charging and permit the addition of distilled water which is lost by evaporation and during charging.

BATTERY RATINGS

Storage batteries are rated for voltage and ampere-hour capacity. When each cell of a lead-acid storage battery is rated at 2.0 volts, with three cells, the total voltage of the battery is 6.0 volts. Higher voltage batteries usually contain more cells.

Fig. 7-1 Storage battery.

The current supplied by a storage battery depends on its voltage, physical condition, and the resistance of the load circuit.

The maximum current output is limited by the internal resistance of the cell. This resistance is determined by the condition of the electrolyte, the size of the plates, and the number of plates. Generally speaking, a larger cell is capable of supplying a higher current than a smaller cell. The voltage of a cell, however, is not dependent on the size or number of plates.

AMPERE-HOUR RATING

The time required to discharge a storage battery at a given load current is determined by its ampere-hour capacity. The *ampere-hour* rating is a measure of the total electrical energy which the battery can deliver. The ampere-hour rating is a function of the size and number of plates in a battery. In general, a large battery has a high ampere-hour rating.

A battery rated at 100 ampere-hours will completely discharge in 100 hours at a rate of one ampere per hour, or in 50 hours at a rate of two amperes per hour. The number of hours a battery will last at a given load current can be determined from the following formula for ampere-hour capacity,

$$\text{hours} = \frac{\text{ampere-hours}}{\text{amperes}}$$

For example, how long will a fully charged battery deliver 10 amperes if it is rated at 60 ampere-hours?

$$\text{hours} = \frac{\text{ampere-hours}}{\text{amperes}}$$

$$\text{hours} = \frac{60}{10}$$

$$\text{hours} = 6$$

STATE OF CHARGE

It is poor practice to discharge a battery completely before recharging it. A battery should be charged whenever its condition drops below the normal value. The condition of a battery, referred to as its *state of charge,* is measured by taking a reading of its *specific gravity* with a battery hydrometer. The apprentice should have at least a general knowledge of the meaning of specific gravity to test a storage battery.

Specific Gravity

Specific gravity is the ratio of the weight of a volume of substance to the weight of an equal volume of fresh water. The equation that expresses this statement is:

$$\text{Specific Gravity} = \frac{\text{Weight of a substance}}{\text{Weight of an equal volume of fresh water}}$$

For example, a pint of concentrated sulfuric acid weighs approximately 1.84 pounds. A pint of fresh water weighs approximately 1 pound. The specific gravity is determined as follows:

$$\text{Specific Gravity} = \frac{1.84}{1} = 1.84$$

The important part of a *hydrometer,* the instrument used to measure specific gravity, is the float on which a scale of specific gravities is marked. The float sinks in a liquid to a certain level, depending on the specific gravity. The lower the float sinks, the smaller the value of specific gravity. Therefore, in sulfuric acid, the float will sink until the surface of the liquid is at the 1.84 value.

LEAD-CELL ACTION

The liquid electrolyte in a fully charged storage cell is made up of sulfuric acid and water. When a cell discharges, acid leaves the electrolyte and combines with lead on the plates. As a result, the electrolyte becomes less dense and lower in specific gravity.

The specific gravity of a fully charged cell is approximately 1.28. A normally discharged cell has a specific gravity of 1.15. The decimal point is commonly omitted for convenience. Therefore, the numbers above are usually referred to as 1280 and 1150.

BATTERY TESTING

The state of charge is usually measured by opening a vent plug of the cell and drawing electrolyte into the barrel of the hydrometer. The scale reading on the float at the level of the liquid is the specific gravity reading.

A battery can also be tested with a high-current discharge tester. This is simply an ammeter combined with a load circuit. A high reading indicates a fully charged battery, and a low reading indicates a need for charging. The ammeter in this instrument is usually calibrated in terms of the state of charge.

BATTERY CHARGING

It is recommended that a battery be charged once a month or whenever its specific gravity falls to 1150. Low specific gravity readings result from normal discharge or because the battery has been allowed to remain inactive. Completely discharged batteries must be recharged immediately. A permanent reduction of the ampere-hour capacity, due to hardening of chemicals on both electrodes, results from letting the battery stand discharged.

Charging Rate

The normal charging rate, in general, is the current specified on the nameplate or in the manufacturer's literature. For a quick charge, a current value a few times higher than the normal value may be used if the temperature of the electrolyte is kept below 110°F.

Charging Current

Either dc or pulsating dc may be used to charge batteries. In either case, the direction of the charging current (electron movement) must be opposite to the current during discharge as shown in figure 7-2. A charging current is produced by connecting the battery to a charger with electrical polarities as marked in figure 7-2B.

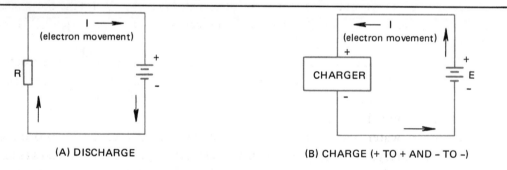

(A) DISCHARGE   **(B) CHARGE (+ TO + AND – TO –)**

Fig. 7-2 Battery current.

The charging rate depends on the voltage difference between the battery voltage and the voltage of the charging source. In all instances, the voltage of the charger must be greater than the total battery voltage. If the charger voltage were lower than the battery voltage, the battery would discharge by driving electrons through the charger.

In engine-driven vehicles, batteries are charged by a dc generator or an alternator that is mounted in the vehicle. When a high-voltage dc supply is available, batteries may be charged directly from the source by the use of suitable current-limiting devices. When an ac supply must be used, the voltage must be rectified; that is, changed to dc before being applied to the battery.

Charging Systems

Devices used to charge batteries operate on the constant-current or constant-potential system. In the constant-current system the charging rate remains the same regardless of battery condition.

In a constant-potential system, the voltage of the charger is held constant at a value slightly above the battery voltage. As the battery charges, its voltage increases slightly, thus reducing the voltage differential between the battery and charger. The result is a high charging rate in the beginning and a low charging rate near the finish; in other words, a tapering charge. This is very desirable since the charging rate is dependent on battery condition.

BATTERY MAINTENANCE

The life of a storage battery depends on the use to which it is put and on the care it receives. With good care it will last several years; with little or no care it may be ruined in a month. The important rules for battery care are as follows:

1. Test storage batteries at least once every week and recharge them at least once each month.

2. If a battery is completely discharged, recharge it immediately.

3. In charging a battery, select a charging rate consistent with the time available for charging. When time is available, use the normal rate indicated on the nameplate.

4. If it is necessary to charge a battery at a very high rate, keep a careful check on the temperature of the electrolyte and *never* let it exceed 110°F. If cells release gas freely, reduce the charging rate to the normal rate.

5. Never try to charge batteries to a definite specific gravity. Maintain the charge until the same specific gravity reading is indicated at three successive half-hour intervals.

6. By the regular addition of distilled water only, maintain the level of the electrolyte above the top of the separators, according to the manufacturer's specifications. Rapid deterioration of a battery will result if the electrolyte level is allowed to remain below the top of the separators.

7. Add distilled water immediately before recharging a battery. In the process of charging a battery, the water in the electrolyte is changed into hydrogen gas and oxygen gas which escape through the vent holes. This water must be restored so that the level of the electrolyte is maintained.

8. **Never use a match to provide light when checking electrolyte level. Hydrogen and oxygen mixed together are highly volatile. The area used for recharging must be well ventilated.**

9. **Never disconnect the leads to a battery while it is on charge. The spark that occurs at the terminals may ignite the gas and cause an explosion.** Many times, a battery is to be charged while permanently mounted in position, such as on an automobile, where the negative terminal may be connected to a frame or an engine. **To reduce the chance of an explosion, the negative lead of the charger should be connected to the frame instead of to the terminal.**

10. Never take a specific gravity reading just after adding distilled water to a battery. Addition of distilled water dilutes the electrolyte and lowers the specific gravity. A reading then would indicate a state of charge below the actual condition of the battery.

11. Avoid spilling electrolyte when testing a battery with a hydrometer.

12. Never add acid or electrolyte to a battery unless it has been definitely determined that some electrolyte has been lost. If it is ever necessary to prepare electrolyte, remember that *acid must be added to water, and must be added slowly.*

13. When placing a battery on charge, do not remove the vent plugs. The plugs will prevent acid spray from reaching the top surface of the battery.

14. Remove deposits that may form on the terminals of a storage battery so that the metal will not be eaten away. The presence of a greenish-white deposit on battery terminals indicates corrosion. Remove this material by thoroughly cleaning the affected parts with a wire brush. Then apply a strong solution of baking soda and water to all corroded parts to neutralize any acid that remains. Wash the battery with fresh water and dry with compressed air or a cloth. Finally, coat the terminals with petroleum jelly or other suitable material.

15. Do not draw a heavy discharge current except for short intervals of time. If high current is needed for a long period, use additional batteries connected in parallel.

16. Test storage batteries more frequently in very cold weather than in warm weather. A discharged battery freezes easily.

ACHIEVEMENT REVIEW

In questions 1–4, complete the statement with a word or phrase to make the statement correct.

1. Batteries are rated in voltage and _____ capacity.

2. Data for use in charging a battery is found on its _____ .

3. The electrical condition of a battery is referred to as its _____ of _____ .

4. The instrument used to determine specific gravity is the _____ .

In items 5–13, select the *best* answer to make each statement true. Place the letter of the answer in the space provided.

5. The state of charge of a battery is measured with   _____
   a. a voltmeter.   d. an ohmmeter.
   b. an ammeter.   e. a thermometer.
   c. a hydrometer.

6. The maximum possible current output of a cell is determined by the   _____
   a. internal resistance of the cell.
   b. link connector.
   c. charging rate.
   d. separator material in the battery.
   e. load resistance.

7. When secondary cells have discharged, they are commonly   _____
   a. recharged
   b. discarded.
   c. allowed to stand idle.
   d. discharged further to 1150.
   e. put on a time tester.

8. When charging a battery at a high rate, the rate must be   _____
    reduced to the normal rate if the
    a. electrolyte temperature exceeds 100°F.
    b. charger voltage is less than the battery voltage.
    c. terminals are not coated.
    d. internal resistance increases.
    e. cells release gas freely.

9. While a cell is being discharged, the electrolyte   _____
    a. becomes more dense.
    b. becomes less dense.
    c. develops a higher specific gravity.
    d. should be replaced.
    e. temperature should be checked.

10. Rapid deterioration of a battery will take place if   _____
    a. it is allowed to remain charged.
    b. the electrolyte level is allowed to remain below the
       top of the separators.
    c. pulsating dc is used to charge it.
    d. it is charged at a high rate.
    e. it is recharged too often.

11. A large storage cell, as compared to a small one, has a   _____
    a. higher voltage.
    b. longer life.
    c. larger current capacity.
    d. lower freezing point.
    e. higher internal resistance.

12. The condition of a battery is determined by the   _____
    a. voltage rating.
    b. ampere-hour rating.
    c. terminal voltage under load.
    d. specific gravity of the electrolyte.
    e. quantity of electrolyte.

13. Storage batteries are rated for ampere-hour capacity and   _____
    a. voltage.   d. energy.
    b. current.   e. internal resistance.
    c. power.

14. Give the equation for determining the time required to fully discharge a com-
    pletely charged 90-ampere-hour, 12-volt battery, if it delivers a constant current
    of 15 amperes to a load.

Electrical Conductors and Wire Sizes

8

OBJECTIVES

After studying this unit, the student will be able to

- describe the factors that determine the resistance of a conductor.
- use the wire gauge tables.

All electrical power is distributed by a system of conductors. The selection and installation of conductors is, therefore, an important practical phase of any electrician's work.

TOTAL CIRCUIT RESISTANCE

It is important to know the factors that contribute to the total resistance of a circuit and the part which conductors contribute to this total.

In any circuit, five factors contribute to the total circuit resistance:
1. The number and type of devices acting as the load circuit.
2. The type of circuit arrangement of these devices.
3. The resistance of switching and control devices.
4. The resistance of conductors carrying power to the devices from the source.
5. The internal resistance of the voltage source.

In general, the first and second factors determine the major portion of the total circuit resistance. Figure 8-1 is an illustration of the effect that grouping of devices has on the total circuit resistance. Notice the different total resistance values.

When the resistance of the load circuit is very low, the resistance of the conductors may become an appreciable part of the total circuit resistance. The large conductors

Fig. 8-1 Total circuit resistance.

used in the starter motor circuit of an automobile are necessary because of the low resistance of the load circuit.

Part of the total voltage applied to a load exists across the conductor. It is always desirable to keep this voltage drop as small as possible. The selection of the proper wire size is often a compromise between the permissible voltage drop and the cost of installing conductors which would yield a lower voltage drop.

CONDUCTOR RESISTANCE

The resistance of a conductor depends on four separate factors:

- the type of material used for the conductor, such as copper or aluminum.
- the length of the conductor.
- the cross-sectional area of the conductor.
- the temperature of the conductor.

Material

Silver is the best conductor of electricity but it is seldom used because of its cost. Copper is almost as good a conductor as silver, is relatively inexpensive, and is adequate for most types of wiring. Aluminum is used where lightness of weight is an important factor. Alloys of copper and various other metals are widely used in the construction of heating elements and for other electrical devices.

Length

The resistance of any conductor is directly proportional to its length. Two feet of a particular wire have twice as much resistance as a single foot; three feet have three times the resistance of a single foot.

Cross-sectional Area

Cross-sectional area, or CSA, is the area of a section cut through an object. The CSA of a wire is the amount of surface on the end of a wire cut at right angles to the axis of the wire. In figure 8-2, the shaded section is the cross-sectional area.

The larger the conductor, the lower the resistance and the easier it is to pass current. In more precise terms, the resistance of a conductor is inversely proportional to its cross-sectional area.

Ordinarily, the cross-sectional area is expressed in square inches. For wires, however, the circular mil is the standard unit of area. A *circular mil* is the CSA of a wire 1/1 000 (0.001) inch in diameter as shown in figure 8-3. The length of a piece of wire one circular mil in

CROSS-SECTIONAL AREA
(CSA)

Fig. 8-2 Cross-sectional area .

0.001″

1′ 0″

AREA = 1 CIRCULAR MIL

Fig. 8-3 A mil-foot.

cross-sectional area and one foot long is called a *mil-foot*. At 68°F, the resistance of a mil-foot of copper wire is approximately 10.4 ohms.

The CSA of any conductor, expressed in circular mils, can be found by determining the diameter of the wire in thousandths of an inch (mils) and squaring this number.

Find the CSA in circular mils of a wire 1/100 inch in diameter.

$$1/100 \text{ inch} = 0.010 \text{ inch}$$
$$0.010 \text{ inch} = 10 \text{ mils}$$
$$(10)^2 = 100 \text{ circular mils}$$

If the CSA of a wire is known in circular mils, the diameter can be determined easily.

For example, find the diameter in inches of a wire with a CSA of 100 circular mils.

$$\sqrt{100} = 10 \text{ mils}$$
$$10 \text{ mils} = 0.010 \text{ inch} = 1/100 \text{ inch}$$

Conductor sizes are expressed by numbers in the American Wire Gauge Table, figure 8-4, page 42. Observe from this table that 1 000 feet of No. 10 wire has a resistance of almost one ohm. Also note that for every third gauge (for example, from 3 to 6) the wire halves in cross-sectional area and doubles in resistance.

It has been found that the resistance of a wire is directly proportional to its length and inversely proportional to its cross-sectional area. This information can be written as a formula.

$$R = \frac{KL}{CSA}$$

Where R = resistance of wire in ohms
      K = resistance per mil-foot of the wire (10.4 ohms for copper)
      L = length in feet
    CSA = cross-sectional area in circular mils

For example, find the resistance of a No. 14 gauge copper wire, 175 feet long, at 68°F.

From figure 8-4, No. 14 gauge wire has a circular mil area of approximately 4 107 circular mils, so that

$$R = \frac{10.4 \times 175}{4\ 107} = 0.443 \text{ ohm}$$

Temperature

The resistance of a circuit or conductor is usually constant and does not depend upon either current or voltage. However, if the current is excessive, the temperature may rise and cause an increase in resistance. When the filament of an incandescent lamp is hot, it has a much higher resistance than when it is cold. Carbon is an exception to this statement because its resistance decreases as the temperature increases. Certain alloys, such as manganin, maintain nearly constant resistance through wide variations in temperature.

B & S Gauge Number	Diameter in Mils	Area in Circular Mils	Ohms per 1 000 ft (ohms per 100 m) Copper 68°F (20°C)
0000	460.00	211 600	0.048 93
000	409.64	167 810	.061 70
00	364.80	133 080	.077 80
0	324.86	105 530	.098 11
1	289.30	83 694	.123 9
2	257.63	66 373	.156 0
3	229.42	52 634	.196 7
4	204.31	41 742	.248 0
5	181.94	33 102	.312 8
6	162.02	26 250	.394 4
7	144.28	20 816	.497 3
8	129.49	16 509	.627 1
9	114.43	13 094	.790 8
10	101.89	10 381	.997 2
11	90.742	8 234.0	1.257
12	80.808	6 529.9	1.586
13	71.961	5 178.4	1.999
14	64.084	4 106.8	2.521
15	57.068	3 256.7	3.179
16	50.820	2 582.9	4.009
17	45.257	2 048.2	5.055
18	40.303	1 624.3	6.374
19	35.890	1 288.1	8.038
20	31.961	1 021.5	10.14
21	28.462	810.10	12.78
22	25.347	642.40	16.12
23	22.571	590.45	20.32
24	20.100	404.01	25.63
25	17.900	320.40	32.31
26	15.940	254.10	40.75
27	14.195	201.50	51.38
28	12.641	159.79	64.79
29	11.257	126.72	81.70
30	10.025	100.50	103.0
31	8.928	79.70	129.0
32	7.950	63.21	163.8
33	7.080	50.13	206.6
34	6.305	39.75	260.5
35	5.615	31.52	328.4
36	5.000	25.00	414.2
37	4.453	19.82	522.2
38	3.965	15.72	658.5
39	3.531	12.47	830.4
40	3.145	9.89	1 047

Fig. 8-4 American wire gauge table for solid copper wire.

Current through a conductor causes the production of heat within the conductor. The resultant rise in temperature sets a limit on the amount of current that can be passed through a conductor. The resistance of a conductor also varies if there is a variation in the temperature of the environment surrounding the conductor. For copper, if the temperature increases, the resistance will increase.

The heat produced within a conductor radiates into space or is conducted away by materials in contact with the wire. If heat is produced faster than it is dissipated, the conductor may melt. For obvious reasons of safety, the National Electrical Code sets definite limits on the amount of current which a conductor is permitted to carry. Since both the type and thickness of insulation are factors in retarding the dissipation of heat, they must be considered in selecting conductors. The National Electrical Code specifies the current-carrying capacities of various wire sizes with different types of insulation when a specific number of conductors are installed in a raceway or cable. Therefore, the current-carrying capacity of each conductor depends on the number of wires present in a raceway or cable.

In general, all conductors must be protected in accordance with their allowable current-carrying capacities. The Code should be consulted for further information regarding specific installations.

ACHIEVEMENT REVIEW

Select the *best* answer for items 1–10 to make each statement true, and place the letter for the answer in the space provided.

1. The resistance of a copper conductor is inversely proportional   _____
   to the
   a. cross-sectional area.   d. current through
   b. gauge numbers.   the conductor.
   c. surrounding temperature.   e. length.

2. A conductor has a resistance of 6 ohms. A conductor of the same   _____
   material and length but twice the CSA will have a resistance of
   a. 1/3 ohm.   d. 6 ohms.
   b. 2 ohms.   e. 12 ohms.
   c. 3 ohms.

3. A conductor has a resistance of 10 ohms. A conductor of the   _____
   same material but twice the diameter will have a resistance of
   a. 1/5 ohm.   d. 10 ohms.
   b. 2.5 ohms.   e. 20 ohms.
   c. 5 ohms.

4. A conductor has a resistance of 12 ohms. A second conductor   _____
   with the same material and cross-sectional area is four times as
   long, and has a resistance of
   a. 1/3 ohm.   d. 12 ohms.
   b. 3 ohms.   e. 48 ohms.
   c. 6 ohms.

5. The CSA of wires is measured in   _____
   a. mils.   d. circular feet.
   b. circular mils.   e. square inches.
   c. mil-feet.

6. By mathematically squaring the number of mils in the diameter   _____
   of a wire, the result is the
   a. cross-sectional area.   d. mil-feet.
   b. length.   e. conductance.
   c. resistance.

7. The number of separate conductors in a raceway or cable is a   _____
   major factor in determining the
   a. resistance of each conductor.
   b. voltage of each conductor.
   c. size of the conductors.
   d. current-carrying capacity of the conductors.
   e. temperature at which the conductors will melt.

8. The greatest portion of total circuit resistance is determined by _____
   the type of circuit arrangement and the
   a. internal resistance of the voltage source.
   b. resistance of control devices.
   c. load resistance.
   d. resistance of the wires.
   e. resistance of switching devices.

9. The kind of insulation on a conductor partially determines the   _____
   conductor's
   a. current-carrying capacity.   d. resistance.
   b. cross-sectional area.   e. material.
   c. gauge.

10. The resistance of an aluminum wire is directly proportional to the _____
    a. CSA.   d. number of circuit
    b. temperature of the wire.   control devices.
    c. circular mils.   e. source voltage.

11. Name the five factors which determine the total resistance of a circuit.

    _____

    _____

    _____

    _____

    _____

12. Name four factors which affect the resistance of a conductor.

_____

_____

13. Find the CSA of a wire with a diameter of 17/1 000 inch.

_____

14. Find the diameter of a wire with a CSA of 311 000 circular mils.

_____

15. A one-foot piece of solid copper wire at 68°F has a resistance of 0.129 ohms. From the Wire Gauge Table, calculate the gauge of the wire.

_____

_____

_____

_____

_____

16. Determine the resistance of a No. 12 gauge copper wire at 68°F if it is 1 883 feet in length. _____

_____

_____

_____

_____

Voltage Drop Across Conductors

9

OBJECTIVES

After studying this unit, the student will be able to

- discuss the principles of voltage drop across conductors.
- demonstrate the problem-solving techniques involved in the selection of conductors.

The effects of voltage drop across conductors can be observed each time the lights in a home dim as a toaster or electric iron is connected. This effect is produced when a low-resistance device is connected directly to the line or feeder. Because of this annoying effect, power companies limit the power ratings of devices which are connected directly to a line, and specify current-limiting controllers for use with motors and other high-current loads. Assuming that the power supply has sufficient electrical capacity, this dimming effect can be reduced by using large conductors with low resistance. The selection of a conductor is usually a compromise between cost and the permissible voltage drop.

When a conductor carries current, a certain percentage of the voltage applied to the total circuit appears across the conductors supplying the load. This voltage, known as the line drop, potential drop, or voltage drop along the line, is due to the resistance of the line conductors. The actual voltage applied to the load is equal to the total voltage minus the line drop.

According to Ohm's Law, the voltage required to produce a current through a given resistance is equal to the product of the desired current and the resistance. In the case of line conductors, the resistance is that of both conductors.

Figure 9-1 shows a dc generator supplying 50 amperes to a motor located 250 feet away. The conductors are No. 4 copper wire. Find the voltage applied to the motor if the generator operates at 257 volts.

Fig. 9-1 Voltage drop example No. 1.

A simple formula can be used to determine the voltage drop or the wire size.

$$V_d = \frac{KIL}{CSA} \qquad or \qquad CSA = \frac{KIL}{V_d}$$

Where V_d = Permissible voltage drop in volts

K = 10.4 ohms per mil-foot (for copper)

I = Current in amperes

L = Total length of circuit in feet

CSA = Cross-sectional area in circular mils
(see Wire Gauge Table, figure 8-4)

Thus, for the example of figure 9-1,

$$V_d = \frac{10.4 \times 50 \times 500}{41\ 742} = 6.22 \text{ volts (total voltage dropped across the wires)}$$

$$V_M = E_G - V_d = 257 - 6.22 = 250.78 \text{ volts across the motor}$$

Figure 9-2 shows a motor operating from a 220-volt dc source and drawing 50 amperes at full load. If a 2 percent drop in line voltage is permitted, find the smallest size conductor to use in a line 100 feet in length.

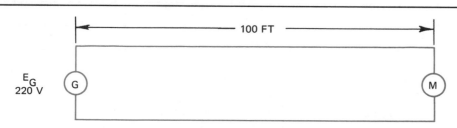

E_G 220 V   G   M

100 FT

Fig. 9-2 Voltage drop example No. 2.

The total line drop is 2 percent of 220 or 4.4 volts.

$$CSA = \frac{KIL}{V_d} = \frac{10.4 \times 50 \times 200}{4.4} = 23\ 636 \text{ circular mils}$$

From the Wire Gauge Table, the proper size wire is No. 6 which has an area of 26 250 circular mils. Always use a wire size equal to or larger than the answer obtained from the formula.

For all practical purposes, voltage drop is not affected by the insulation on a conductor. The higher temperature insulations will carry their rated current according to the tables in the National Electrical Code, but the voltage drop may not be kept to a minimum. When long distances are involved, conductor sizes are usually determined by first considering the voltage drop and then selecting an insulation suitable for the temperature and location encountered.

ACHIEVEMENT REVIEW

1. A dc motor draws 100 amperes at full load from a 220-volt dc source 200 feet away. If a 3 percent line voltage loss is permissible, find the wire size to be used for the line conductors.

_____

_____

_____

2. A 110-volt dc source supplies 25 amperes to a load circuit 500 feet away. If two No. 10 wires are used for line conductors, find:
   a. The line drop
   b. The voltage at the load

_____

_____

_____

3. Determine the proper size conductors necessary to supply a parallel group of five 300-watt lamps which are located 75 feet from a panelboard. The voltage at the panelboard is 120 volts. Permissible line drop is 1.2 volts.

_____

_____

_____

_____

   In items 4–9, select the _best_ answer that will make each item a true statement. Place the letter of the answer in the space provided.

4. The annoying effect of dimming lights in a home sometimes occurs   _____
   when a device is connected to the line and it has a value of resistance that is
   a. high.
   b. low.
   c. medium.
   d. higher than the resistance of other devices on the line.
   e. equal to the resistance of other devices on the line.

5. For a small line-voltage drop, the resistance of line conductors   _____
   should be
   a. a small percentage of the total circuit resistance.
   b. a large percentage of the total circuit resistance.
   c. equal to the load resistance.
   d. equal to the combination of load resistance and source
      resistance.
   e. equal to the source resistance.

6. When a permitted percentage of line voltage drop is specified in a   _____
    problem, the value is computed directly from the
    a. load voltage.
    b. source voltage.
    c. voltage across control devices.
    d. voltage across devices that make up the load resistance.
    e. wire tables.

7. The voltage that is applied to a load is equal to   _____
    a. the source voltage.
    b. the line voltage drop minus the source voltage.
    c. the line voltage.
    d. the line voltage drop.
    e. the source voltage minus the line voltage drop.

8. Line voltage drop is inversely proportional to the   _____
    a. length of the circuit.
    b. current through the load.
    c. cross-sectional area of the conductor.
    d. current-carrying capacity of the conductor.
    e. current through the conductor.

9. Two important general factors that must be considered when   _____
    selecting conductors are permissible voltage drop and
    a. cross-sectional area.
    b. cost.
    c. resistance.
    d. the load devices.
    e. whether the conductors should be silver or copper.

Summary Review
10 of Units 1-9

OBJECTIVE

- To evaluate the knowledge and understanding acquired in the study of the previous nine units.

In items 1–10, insert the word or phrase that will make each incomplete statement true.

1. Electrical pressure is measured in _____ .
2. Electrical current is measured in _____ .
3. The symbol for resistance is the Greek letter _____ .
4. An electrical current is the movement of _____ .
5. The symbol for source voltage is the letter _____ .
6. Resistance is measured in _____ .
7. The symbol for current is the letter _____ .
8. Electrical power is measured in _____ .
9. The symbol for electrical power is the letter _____ .
10. Electrical energy is measured in _____ .

For each of the incomplete statements at the left, items 11–20, select the *best* word or phrase from the right to make each statement true. Place the letter of the word or phrase in the space provided.

11. The resistance of a conductor varies directly with its _____
12. Energy per unit time is _____
13. The cross-sectional area of wire is usually expressed in _____
14. The resistance of a wire varies inversely with its _____
15. The best conductor of electricity is _____
16. In wire size tables, the diameter of a wire is expressed in _____
17. A wattmeter is used to measure electrical _____
18. Electrical resistance is measured with an instrument known as _____
19. Higher voltage is obtained by connecting batteries in _____
20. In charging a battery, it should be kept on charge until there is no further increase in

a. copper.
b. circular mils.
c. power.
d. specific gravity.
e. an ohmmeter.
f. length.
g. cross-sectional area.
h. parallel.
i. series.
j. mils.
k. energy.
l. gold.
m. silver.
n. a voltmeter.
o. a watthour meter.
p. series-parallel.
q. voltage.
r. mil-feet.
s. current.

Select the *best* answer to make statements 21–52 true, and place the corresponding letter in the space provided.

21. To operate properly, electrical devices connected in a series circuit must have the same rating with respect to (a) voltage (b) current (c) resistance (d) power (e) insulation.   _____

22. To operate properly, electrical devices in a parallel circuit must have the same rating with respect to (a) voltage (b) current (c) resistance (d) power (e) energy.   _____

23. The major part of the resistance in a correctly wired electrical circuit is usually found in the (a) connecting wires (b) devices in a circuit (c) power supply (d) measuring instruments (e) closed switches.   _____

Refer to figure 10-1 for problems 24–30.

Fig. 10-1.

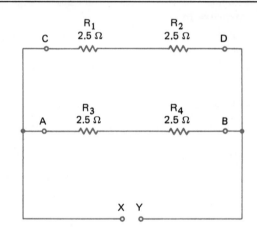

24. Figure 10-1 is a (a) series circuit (b) parallel circuit (c) series-parallel circuit (d) shunt circuit (e) reciprocal circuit.   _____

25. The resistance of branch A-B is (a) 2.5 ohms (b) 5 ohms (c) 20 ohms (d) 0 ohms (e) infinity.   _____

26. If an ohmmeter is connected across R_1, it will read (a) 2.5 ohms (b) 5 ohms (c) 7.5 ohms (d) 10 ohms (e) less than 2.5 ohms. (Note: Consider the entire circuit, not just R_1.)   _____

27. If R_1 is shorted, the total resistance across points X and Y will (a) increase (b) decrease (c) remain constant (d) equal 2.5 ohms.   _____

28. If R_1 develops an "open," the resistance of branch C-D will (a) increase (b) decrease (c) remain constant (d) change to zero.   _____

29. An ohmmeter is connected across points X and Y. It has a reading of infinity. This indicates that (a) one resistor is open (b) R_1 or R_2 is open (c) R_3 or R_4 is open (d) one resistor in each branch is open (e) R_1 and R_2 are both open.   _____

30. An ohmmeter connected across points X and Y has a reading of _____ zero. This indicates that (a) R_1 and R_2 are shorted out (b) one resistor in each branch is shorted out (c) R_2 and R_4 are shorted out (d) R_1 is shorted and R_4 is open.

Refer to figure 10-2 for problems 31–35.

Fig. 10-2.

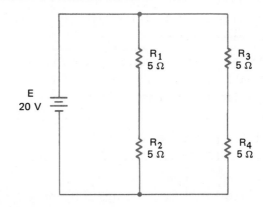

31. The voltage across R_3 is (a) 2 volts (b) 4 volts (c) 5 volts _____ (d) 8 volts (e) 10 volts.

32. The current through R_2 is (a) 1 ampere (b) 2 amperes _____ (c) 4 amperes (d) 8 amperes (e) 16 amperes.

33. The total current in the circuit is (a) 4 amperes (b) 10 amperes _____ (c) 15 amperes (d) 20 amperes (e) 25 amperes.

34. If a voltmeter connected across R_1 has a reading of 20 volts, _____ it means that (a) R_1 is open (b) R_3 is shorted (c) R_1 is shorted (d) R_2 is open (e) R_2 and R_4 must be open.

35. A voltmeter connected across R_2 has a reading of zero. This _____ means (a) R_1 is shorted (b) R_2 is shorted (c) R_3 is open (d) R_2 is open (e) no conclusion can be made.

36. In figure 10-3, the total circuit resistance is (a) 1/4 ohm _____ (b) 1 ohm (c) 4 ohms (d) 8 ohms (e) 16 ohms.

Fig. 10-3.

37. In figure 10-4, the total resistance across points A and B is _____
    (a) 5 ohms (b) 10 ohms (c) 15 ohms (d) 20 ohms
    (e) 25 ohms (f) not obtainable.

Fig. 10-4.

38. A 40-volt power supply is connected across points A and B _____
    of figure 10-4. The current through one of the 10-ohm resistors is
    (a) 1.6 amperes (b) 2 amperes (c) 4 amperes (d) 8 amperes
    (e) 16 amperes.

39. With respect to one another, the three lamps in figure 10-5 are _____
    connected in (a) series (b) parallel (c) shunt (d) series-parallel
    (e) a Norton circuit.

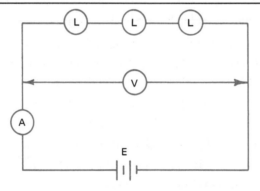

Fig. 10-5.

40. In figure 10-5, the voltmeter is (a) in series with the lamps _____
    (b) in parallel with the lamps (c) in series with the battery
    (d) in parallel with the ammeter (e) in parallel with the center
    lamp.

41. The ammeter in figure 10-5 is (a) in series with the battery _____
    (b) in parallel with the battery (c) in parallel with the voltmeter
    (d) in parallel with the lamps (e) in a short circuit.

42. Four lamps, with equal resistance values, are connected in _____
    parallel to a 120-volt dc power supply. The voltage across each
    lamp is (a) 2.5 volts (b) 25 volts (c) 120 volts (d) 400 volts
    (e) not obtainable.

43. Five lamps, with equal resistance values, are connected in series to _____
    a 125-volt source. The voltage across each lamp is (a) 5 volts
    (b) 25 volts (c) 125 volts (d) 625 volts (e) not obtainable.

44. Five resistors of equal value are connected in parallel to a 125-volt dc supply. If the total current is 5 amperes, the current through one of the resistors is (a) 1 ampere (b) 5 amperes (c) 10 amperes (d) 15 amperes (e) 25 amperes.   _____

45. Four lamps with unequal resistance values are connected in series to a 117-volt supply. If the total source current is 8 amperes, the current through one of the lamps is (a) 2 amperes (b) 4 amperes (c) 8 amperes (d) 16 amperes (e) 32 amperes.   _____

46. The words "state of charge" refer to (a) the specific gravity of the battery (b) the combined voltage of all cells (c) the number of ampere-hours available for discharge (d) the ampere-hour rating of the battery (e) the voltage rating of the battery.   _____

47. The term "charging rate" refers to the (a) cost of charging the battery (b) number of hours needed to charge the battery (c) the charging current (d) the voltage of the charging source (e) cost of the battery charger.   _____

48. A large storage cell, as compared to a small one, has a (a) higher voltage (b) longer life (c) larger ampere-hour rating (d) lower freezing point (e) higher internal resistance.   _____

49. A battery may be charged at a high charging rate if the (a) rate is kept below 150 amperes (b) charging time is below three hours (c) voltage does not exceed 7.5 volts (d) electrolyte temperature is kept below 110°F (e) water level is proper.   _____

50. A storage battery should be charged until (a) the voltage reaches 6 volts (b) the cells begin to gas (c) the specific gravity reaches 1 300 (d) the temperature reaches 110°F (e) the specific gravity reading stops rising.   _____

51. A battery rated at 120 ampere-hours will deliver a current of 5 amperes for approximately (a) 5 hours (b) 12 hours (c) 24 hours (d) 120 hours (e) 600 hours.   _____

52. The outstanding danger of allowing a battery to remain in a state of discharge is that it will (a) result in a permanent reduction in the ampere-hour capacity (b) require a long time to recharge (c) gas violently when charged (d) become damaged at low temperatures or very high altitudes (e) not come up to full current on charge.   _____

53. What is the resistance of a toaster that draws 10 amperes when connected to a 120-volt circuit?   _____

54. Determine the resistance of 1 500 feet of copper wire which has a diameter of approximately 129 mils (K = 10.4).   _____

55.   Two resistors connected in parallel have a combined   _____
      resistance of 12 ohms. One of them is a 48-ohm resistor.
      What is the resistance of the other resistor?

56.   A 4-ohm, an 8-ohm, and a 12-ohm resistor are connected   _____
      in series. The voltage across the 8-ohm resistor is 80 volts.
      Determine the supply voltage.

57.   Calculate the overall efficiency of a dc motor that draws   _____
      40 amperes from a 115-volt source, and delivers 5 hp
      to a load.

11

Magnets and Magnetic Fields

OBJECTIVES

After studying this unit, the student will be able to

- describe the properties of magnets.
- discuss the basic principles of magnetism.

Much of the present-day electrical equipment functions because of magnetism. Motors and generators operate on the principle of magnetism. It is essential that the student of electricity understands this phenomenon.

MAGNETIC MATERIALS

Iron, and its derivative, steel, can be given the property of attracting other pieces of iron and steel. This property, known as *magnetism,* is possessed to a much lesser degree by nickel, cobalt, and gadolinium. Iron and steel combined with these and other magnetic materials will yield an alloy with much greater magnetic strength.

The magnetic effects of magnets are concentrated at areas called *poles.* These poles are of two types and have been designated as north and south poles because of the fact that a magnet supported freely in air will align its axis in a north-south direction. The end of the magnet which points geographically north is called the north (N) pole, the other end is called the south (S) pole. Although all materials have some degree of a magnetic property, most materials do not have a useful amount of this property and, for all practical purposes, can be called nonmagnetic.

PERMANENT AND TEMPORARY MAGNETS

Hard steel is used for the construction of permanent magnets. Soft steel is easier to magnetize, but will retain a relatively weak degree of magnetization when the magnetizing force is removed. This small amount of magnetism retained by soft steel is known as *residual magnetism* and is both desirable and important in the operation of electrical equipment.

ELECTROMAGNETS

A very powerful temporary magnet can be made by placing a bar of soft steel inside a coil of wire carrying an electrical current. The intense magnetic force created is reduced to a weak residual force as soon as the current is interrupted. An electromagnet also can be used to magnetize magnetic materials by placing the material across the poles of the electromagnet, figure 11-1, or by placing the material inside the coil itself.

MATERIAL TO BE
MAGNETIZED

Fig. 11-1 Magnetic charger.

TO DC CHARGER

MAGNETIC INDUCTION

Magnetic materials also can be magnetized by placing them near a magnet. The magnetism produced in the material by this method is called *induced magnetism.* In the case of soft steel, the effect is only temporary. The magnetism is lost as soon as the magnet is removed.

LAW OF MAGNETS

If two magnets are brought near each other, the following will result:

* like poles repel.
* unlike poles attract.

Figure 11-2 illustrates this law. Two N poles and two S poles repel each other. An N pole and an S pole attract each other.

Fig. 11-2 Like poles repel; unlike poles attract.

REPULSION | REPULSION | ATTRACTION

MAGNETIC FIELDS

Magnets influence one another at a distance without actually making contact. The space around a magnet, through which this invisible force acts, is known as the *magnetic field.* The force itself may be represented by *magnetic lines of force* which are assumed to exist in the space between the poles of the magnet. These invisible lines, collectively referred to as *magnetic flux,* are shown in the space around the bar magnet in figure 11-3. Magnetic lines of force cannot be blocked or insulated, but will pass through, or within, any material.

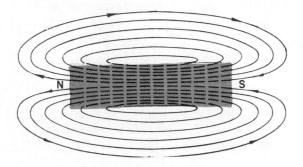

Fig. 11-3 Flux pattern.

Field Strength

The concentration of lines of force is an indication of the magnetic strength at various points in the magnetic field. This concentration, often referred to as the *flux density,* is the number of flux lines in a square inch of cross-sectional area. In other words, as the number of flux lines per cross-sectional area increases, the stronger the magnetic field becomes.

Properties of Magnetic Flux

The following accepted properties of magnetic flux are very useful in explaining the operation of a wide variety of electrical equipment using magnetic circuits.

1. There is no insulator for magnetic flux; it passes through all materials.
2. Lines of force are closed loops passing through the magnet and the space around it.
3. The loops, formed by the lines of force, tend to become larger and increase in length as they develop away from the magnet.
4. Lines of force have direction. They emerge from the north pole and enter the south pole.
5. Lines of force never cross one another.
6. Lines of force concentrate at the poles and develop maximum field strength there.
7. Large numbers of flux lines are easily established in magnetic materials, but are difficult to establish in nonmagnetic materials such as air.

ACHIEVEMENT REVIEW

Select the *best* answer to make each statement true, and place the letter of the answer in the space provided.

1. When a magnetizing force is removed from a material, the kind
   of magnetism that remains is called   _____
   a. strong.   d. electromagnetism.
   b. weak.   e. flux.
   c. residual.

2. The kind of magnet that is made by wrapping a coil of wire
   around a bar of steel is called   _____
   a. a permanent magnet.   c. a transformer.
   b. an electromagnet.   d. a pole magnet.

3. Magnetic lines of force are known as   _____
   a. induction.   c. attracting influences.
   b. poles.   d. flux.

4. Flux density is an indication of   _____
   a. repulsion.   d. field strength.
   b. an electromagnet.   e. a temporary magnet.
   c. a permanent magnet.

5. The magnetism present in a piece of soft steel held near a
   magnet is called   _____
   a. induced magnetism.   d. electromagnetism.
   b. residual magnetism.   e. permanent magnetism.
   c. insulated magnetism.

6. The number of lines of force per cross-sectional area is a
   measure of   _____
   a. magnetic flux density.   d. flux.
   b. magnetic intensity.   e. flux patterns.
   c. the laws of magnets.

7. Magnetic properties are possessed by   _____
   a. iron and steel only.
   b. nickel, cobalt, and gadolinium only.
   c. the materials stated in (a.) and (b.) only.
   d. hard and soft steel only.
   e. all materials.

12 Electromagnetism

OBJECTIVES

After studying this unit, the student will be able to

- discuss the basic principles of electromagnetism.
- demonstrate how to determine the direction of a magnetic field.
- explain how a magnetic field is created in a coil of wire.

Magnetic circuits are employed in generators, alternators, motors, transformers, relays, and many other important electrical machines. In all but a few instances, the magnetizing force is produced by the effects of an electrical current in a coil with an iron core.

CONDUCTOR FLUX

A wire carrying an electrical current exhibits magnetic characteristics. If placed near iron filings it will attract them, as shown in figure 12-1.

Figure 12-2 illustrates the effect which a current-carrying conductor produces on iron filings placed on a surface at right angles to the conductor.

The pattern formed by the iron filings indicates the presence of a circular magnetic field around the conductor. To prove that this field has direction, small magnetic

IRON FILINGS

ELECTRICAL
CONDUCTOR

Fig.12-1   An electrical current is accompanied
by a magnetic field.

Fig. 12-2   Magnetic field around a conductor.

compasses may be placed in the vicinity of the conductor. Figure 12-3 shows that the magnetic compasses, which ordinarily point north and south, will arrange themselves in a circle. Figure 12-3 also shows that the direction of the magnetic flux (conductor flux) depends on the direction of the current. Thus, a magnetic field can be established in either direction by controlling the direction of current in the conductor.

Fig. 12-3 Flux direction about a conductor.

Certain symbols are used to simplify the indication of current direction in a conductor. The *dot* and *cross* method is illustrated in figure 12-4. A *dot* indicates current coming toward the observer; a *cross* indicates current going away from the observer.

Fig. 12-4 Dot-cross method of indicating current direction (direction of electron movement).

The two cross-sectional views in figure 12-5 illustrate the distribution and direction of flux around a current-carrying conductor for both directions of current. Note that the flux density is greatest near the wire and that individual lines of flux are closed loops.

Fig. 12-5 Flux direction — cross-sectional view.

Although a current-carrying conductor has a magnetic field, it does not have poles. A pole is defined in unit 11 as a point where magnetism is concentrated, and as the points where flux lines emerge from a magnet and reenter a magnet. These points do not exist for conductors.

The direction of the flux around a current-carrying conductor can be determined by placing a magnetic compass near the wire. The direction of the_compass N pole defines the direction of the flux at the point where the compass is placed, as shown in figure 12-5.

LEFT-HAND RULE (CONDUCTOR FLUX)

Figure 12-6 illustrates the left-hand rule as it is used to determine the direction of conductor flux. Place the left hand around the conductor with the thumb pointing in the direction of electron movement. You do not have to touch the conductor. The fingers will then wrap around the conductor in the direction of the flux.

Fig. 12-6 Left-hand rule (conductor flux).

Figure 12-7 shows the direction taken by lines of force around a bar magnet. The lines emerge at the north pole and reenter at the south pole.

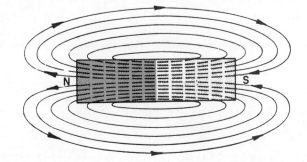

Fig. 12-7 Flux direction in a bar magnet.

A straight current-carrying conductor has no poles. As soon as the same conductor is arranged as a loop, it takes on the polar characteristics of a magnet. By adding additional loops, also called turns, a coil is formed. The magnetic field produced in the vicinity of the coil is shown in figure 12-8, page 64. Note that this field pattern is like that of a bar magnet.

Since the flux emerges at the left side, this end has the properties of a north pole. Flux reenters at the other end of the coil, so this end has the properties of a south pole.

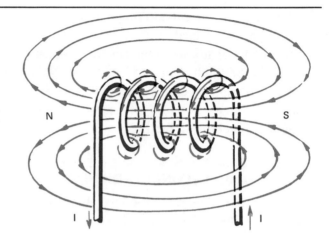

Fig. 12-8 Magnetic polarity of a coil.

LEFT-HAND COIL RULE

Figure 12-9 illustrates the technique of using the left-hand coil rule.

The magnetic polarity of the coil is determined by placing the fingers of the left hand in the direction of current (electron movement) as it exists through the turns of wire. The thumb will point in the direction of the north pole.

Fig. 12-9 Left-hand coil rule.

MAGNETIC STRENGTH

The magnetic strength of a coil depends upon the

- amount of current in the coil.
- number of turns in the coil.
- type of core material used.

Magnetic strength can be expressed in ampere-turns for a given core material. The term *ampere-turn* means the product of current, in amperes, and the number of turns in the coil.

In many cases, materials are inserted in a coil to increase the magnetic strength of the coil. These materials are called *cores*. For a given core material, the magnetic strength will change with a variation in the current and the number of turns, that is, with the ampere-turns.

If a coil has a constant number of turns wrapped around a core, the current is the only factor that can affect magnetic strength. Therefore, the more current there is through a coil, the stronger the magnetic field will be.

If a soft iron bar is inserted in a coil as the core material, a very strong magnetic field is established, as compared to the field produced when an air core is used.

ACHIEVEMENT REVIEW

Select the *best* answer in the following problems to make each statement true. Place your answers in the spaces provided.

1. A straight current-carrying conductor has   _____
   a. two magnetic poles.
   b. one north pole only.
   c. no magnetic poles.
   d. one magnetic pole.
   e. a field similar to a bar magnet.

2. The direction of conductor flux is dependent upon   _____
   a. current magnitude.
   b. current direction.
   c. the compass needle.
   d. the magnitude of voltage applied.
   e. the core material.

3. The magnetic polarity of a coil is determined by   _____
   a. the magnitude of voltage applied.
   b. the current magnitude.
   c. the magnetic strength
   d. the number of turns.
   e. the direction of current.

4. The magnetic strength of a coil depends on   _____
   a. current direction.
   b. the left-hand rule.
   c. flux direction.
   d. current magnitude.
   e. the point where the field emerges.

5. A 20-turn coil, with an air core, carries a current of 2 amperes.   _____
   The magnetic strength of the coil can be increased by
   a. making the turns larger.
   b. inserting an iron core.
   c. decreasing the current.
   d. slightly decreasing the voltage drop across the coil.
   e. reversing the current direction.

6. When iron filings are attracted to a current-carrying conductor,   _____
   this indicates
   a. a magnetic field direction.
   b. a north pole.
   c. the presence of a magnetic field.
   d. the strength of a magnetic field.
   e. the magnitude of current.

7. The properties of a magnet are present in   _____
   a. a loop of wire.
   b. a straight wire.
   c. many turns of wire.
   d. a straight wire carrying current.
   e. a current-carrying loop of wire.

Generation of Electromotive Force

13

OBJECTIVES

After studying this unit, the student will be able to

- discuss the principles involved in the production of an electromotive force.

- explain how voltage is generated due to mechanical motion.

An electromotive force is necessary to produce an electrical current. The production of electrical energy on a large scale cannot be accomplished economically with batteries. Most of the electricity produced today is created through the use of alternators and generators. Both machines operate on the principle of induced voltage.

In figure 13-1, a conductor, which has its ends connected to a sensitive ammeter, is being moved rapidly downward in a magnetic field. As the conductor is moved downward, it cuts lines of magnetic flux. As a result, there is a deflection of the meter needle indicating the presence of an electrical current produced by an induced voltage. It is evident that the motion is responsible for the voltage produced since no current is present if the conductor is held motionless. Furthermore, the meter needle deflects in the opposite direction if the conductor is moved upward through the magnetic field. The direction in which an induced voltage is produced depends on the direction of conductor movement.

CONDUCTOR
MOTION
DOWNWARD

Fig. 13-1 Inducing an electromotive force.

In general, the amount of induced voltage produced in a conductor is directly proportional to the

- strength of the magnetic field.

- length of the conductor in the field.

- speed at which the conductor passes through the field.

- angle at which the conductor passes through the field.

In machines that generate voltage, the strength of the field and the conductor length are fixed quantities. The angle of cutting depends upon the rotation of the conductor. Therefore, the only real variable is conductor speed. As the conductor speed increases, more lines of force are cut per second, and the induced voltage increases in magnitude.

In generators and alternators, powerful electromagnets are used to establish a strong magnetic field. Conductors are mounted on an armature and rotated at high speeds through this field. A large number of conductors can be used so that the individual voltages of all the conductors act in series to produce a greater voltage. In summary then, a high voltage can be created by cutting a powerful magnetic field with a series of conductors moving at high speed.

LEFT-HAND GENERATOR RULE

Figure 13-2 shows how to determine the direction of an induced voltage when the direction of the magnetic field and the direction of conductor motion are known. The induced voltage creates a current that has a direction the same as that of the induced voltage.

Fig. 13-2 Left-hand generator rule.

This method is known as the Left-hand Generator Rule. Position the thumb, first finger, and middle finger of the left hand at right angles to one another. If the hand is placed with the thumb in the direction of conductor motion, and the first finger in the direction of the magnetic field, then the middle finger will point in the direction of the induced current (electron direction).

THE AC GENERATOR

The essential parts of a generator are shown in figure 13-3. A single conductor loop is placed so that it can be rotated in the space between two opposite poles of an electromagnet. To simplify the explanation, one side of the loop is shown in black and the other side is in white. To use the induced voltage in an external circuit, each end of the loop is connected to a slip ring. The external circuit is connected to these rings by means of a brush pressing against each ring. In other words, a complete electrical circuit is provided through the sliding contacts at the slip rings.

Assume that the loop is forced to rotate clockwise in the magnetic field. For the position shown in figure 13-3, the conductors which form the sides of the loop are moving parallel to the lines of force. At this instant no flux is cut by the conductors; therefore, no voltage is generated.

Fig. 13-3 Rotating loop – position 1.

As the loop is rotated it reaches the position shown in figure 13-4. Both sides of the loop now cut flux but in opposite directions. An application of the Left-hand Generator Rule shows that voltage is induced in opposite directions on opposite sides of the loop. This means, however, that in the loop as a whole, the voltages are in the same direction. Carefully note the direction of the current in the external circuit.

Fig. 13-4 Rotating loop – position 2.

As the loop reaches the position shown in figure 13-5, one-half revolution has been completed and both sides of the loop are again moving parallel to the magnetic flux. At this instant no voltage is generated and there is no current in any part of the loop or external circuit.

Fig. 13-5 Rotating loop – position 3.

As the loop is rotated further, it reaches the position shown in figure 13-6. In this position, it has completed three-quarters of one revolution. By applying the Left-hand Generator Rule, the current in the black and white sections of the loop can be determined. Note carefully that the current in both the loop and the external circuit is reversed from that indicated in position 2.

Fig. 13-6 Rotating loop – position 4.

One-quarter revolution later, the loop has reached its original position and the voltage and current again are zero.

Three important facts about the rotating loop must be emphasized.
1. The induced voltage in the loop reverses in direction twice each revolution.
2. An alternating current which reverses itself twice each revolution is present in the external circuit.
3. The voltage and the resulting current are pulsating.

A graph illustrating the variations of induced voltage (emf) for one full revolution of the loop is shown in figure 13-7. The maximum voltage is created whenever

Fig. 13-7 Alternating EMF.

the loop cuts flux at the fastest rate, or when the conductor is moving perpendicular to the lines of force. The part of the graph below the horizontal axis indicates voltage in the opposite direction.

DC GENERATOR

The single loop rotating in a magnetic field can be used to supply a direct current to an external load circuit by means of a simple device known as a *commutator.*

Figure 13-8 illustrates a single loop whose conductors terminate at a commutator consisting of a ring split lengthwise into two separate segments. Because the loop is to be rotated, a sliding contact is necessary to bring current to the load circuit. Two brushes, connected to the load circuit leads, rest against these commutator segments.

Fig. 13-8 Rotating loop — position 1.

Assume that the loop is rotated in a clockwise direction. In figure 13-8, the loop is in a vertical position and no voltage or current is present in any part of the circuit.

In position 2, figure 13-9, page 72, the loop sides are cutting flux. The induced voltage in the loop produces a current from the white wire, to the white segment, to the white brush, and to the load circuit. The black brush is the positive terminal and the white brush is the negative terminal of the generator. Note the direction of the current in the load circuit, from left to right through the meter.

Fig. 13-9 Rotating loop – position 2.

In position 3, figure 13-10, the loop is now in a vertical position again and no voltage or current exists in any part of the circuit.

Fig. 13-10 Rotating loop – position 3.

In position 4, figure 13-11, the induced current has reversed in both sides of the loop. Note, however, that current from the black wire passes to the load circuit by way of the white brush. Observe carefully that the white brush is still negative, and that the current remains in the same direction in the load circuit.

Fig. 13-11 Rotating loop – position 4.

Three important facts concerning this circuit must be emphasized:

1. The induced voltage in the loop reverses itself twice each revolution.
2. The induced voltage and the resulting current are pulsating in character.
3. Although the current in the loop is ac, a direct current exists in the load circuit.

A graph of the voltage developed across the brushes of a single loop rotated in a magnetic field for one complete revolution is shown in figure 13-12.

Fig. 13-12 EMF from a single loop.

The output of a single-loop generator is too small and pulsating for any practical use. Commercial generators use many loops mounted on the rotating member. This has the effect of increasing the voltage and reducing the fluctuations in voltage output.

Many factors determine the voltage output of a generator. The factors are, (1) number of poles, (2) flux per pole, (3) number of conductors on the armature, and (4) speed of the armature. An in-depth study of each of these factors is required for a complete understanding of generator operation.

ACHIEVEMENT REVIEW

In the following problems, select the *best* answer to make the statement true, and place the letter of the answer in the space provided.

1. The direction of induced voltage in a conductor can be changed by   _____
   a. increasing the field strength.   d. decreasing conductor size.
   b. reversing the field direction.   e. reversing meter connections.
   c. increasing conductor length.

2. Direct current can be supplied to a load, by a loop of wire rotating   _____
   through a field, with the use of
   a. slip rings.   d. a commutator.
   b. electromagnets.   e. a conductor.
   c. brushes.

3. Induced voltage can be increased in magnitude by   _____
   a. increasing the number of lines cut per second.
   b. using a commutator.
   c. using slip rings.
   d. decreasing conductor length.
   e. properly applying the Left-hand Generator Rule.

4. The induced voltage in a single loop reverses   _____
   a. once each revolution.
   b. once each half revolution.
   c. twice each half revolution.
   d. twice each revolution.
   e. three times each revolution.

5. Maximum voltage is induced in a single loop when the sides of the   _____
   loop are passing
   a. perpendicular to the lines of force.
   b. parallel to the lines of force.
   c. not quite perpendicular to the lines of force.
   d. at a slow rate of speed.
   e. in front of the north pole-face.

6. When a commutator is used on a single loop, the voltage at the   _____
   brushes has a
   a. very large magnitude.   d. zero value.
   b. changing polarity.   e. constant value.
   c. constant polarity.

7. The Left-hand Generator Rule is usually used to determine   _____
   a. conductor speed.
   b. rotational direction.
   c. field direction.
   d. current direction.
   e. magnetic field strength.

14

Direct-Current Motor Principles

OBJECTIVES

After studying this unit, the student will be able to

- determine the direction of movement of a current-carrying conductor in a magnetic field.

- discuss the basic principles of dc motors.

A large part of the energy used worldwide is created through hydroelectric installations and the burning of coal and oil. This potential energy is in mechanical form and cannot be distributed as such at distances far from its source. If this energy is converted to electrical energy, the problem of distribution is solved. It is necessary, however, to use electric motors to change the energy back to a mechanical form at the point of application.

A simple conversion of electrical energy to mechanical energy is shown in figure 14-1. Figure 14-1A shows a uniform magnetic field in which a conductor, carrying no current, is placed. In figure 14-1B, the field is removed and a current is created in the conductor due to an external voltage source. Notice the field about the conductor that is created by the current (electrons) passing out of the paper. Figure 14-1C illustrates the resultant magnetic field that exists when the magnetic field is added. Above the conductor, the field produced by the current acts in an additive manner with the field created by the poles. Below the conductor, the conductor field acts in opposition to the pole field. The addition of the fields above the conductor, with the reduction of the field below the conductor, causes the conductor to move in a downward direction. In figure 14-1D, the current (electron) direction in the conductor is reversed, going into the page, so that the lines are additive below the conductor, and in opposition above the conductor. In this case, the conductor direction is upward. The conductor movement shown in figure 14-1 is the basic principle which governs the action of a motor.

Fig. 14-1 Force on a current-carrying conductor.

Before this principle can be applied to an actual motor, a rule must be formulated so that the direction of conductor motion can be determined when the direction of the current (electron movement) is known.

RIGHT-HAND MOTOR RULE

The Right-hand Motor Rule is explained by placing the thumb, first finger, and middle finger of the right hand at angles to one another. If the first finger is pointed in the direction of field flux, and the middle finger is in the direction of conductor current (electron direction), then the thumb points in the direction of conductor motion, figure 14-2.

Figure 14-3 shows a single-loop armature placed in the magnetic field between two permanent magnets. The flux established by the permanent magnet is called the field flux.

A current introduced into the loop through the brushes and commutator produces flux around all parts of this loop. This flux is called conductor flux, (described in unit 12). On the right-hand side of the loop, an application of the Right-hand Motor Rule shows that this loop is forced downward. On the left-hand side of the loop, the conditions are reversed and the loop side is forced upward. If this loop is mounted on a shaft and is free to rotate, motion in a clockwise direction results.

In figure 14-4, the loop has reached a vertical position and the brushes rest on the insulated spacer between the commutator segments. No current exists in the loop and no force is present to continue the rotation at this neutral position. The loop, however, has momentum due to the preceding one-quarter revolution and thus passes through this neutral position.

Fig. 14-2 Right-hand Motor Rule.

Fig. 14-3 Single-loop armature — position 1.

Fig. 14-4 Single-loop armature — position 2.

In figure 14-5, the armature continues its movement so that the commutator segments interchange their positions on the brushes and current reverses in the loop. Thus, there is a reversal of conductor flux direction on both the black and white sections of the loop. This means that as each side of the loop passes a pole, the current in the loop is always in the same direction with respect to that pole. As a result, the rotation of the loop is maintained in one direction.

The amount of *torque,* or turning force, developed by this single loop is directly dependent upon the strengths of the field flux and the conductor flux. To strengthen the field flux, it is customary to use electromagnets for the field poles of a motor. To strengthen the conductor flux, the current in the wire must be increased. The maximum turning force is developed when the loop is in a horizontal position; the minimum force results when it is in a vertical position.

The graph of the torque developed by a single-loop armature over a period of one full revolution is shown in figure 14-6. Note that there are two positions of maximum torque and two positions of minimum torque.

A single-loop armature has little practical value for commercial motor applications. The torque applied to the motor shaft is weak and pulsating even with an electromagnetic field.

Fig. 14-5 Single-loop armature – position 3.

Fig. 14-6 Torque graph single-loop armature.

Fig. 14-7 Torque graph double-loop armature.

The undesirable pulsations in torque of a single-loop armature can be eliminated by adding additional loops and the necessary commutator segments. Figure 14-7 shows the torque graph of a double-loop armature. Although the torque is still pulsating, there is a noticeable reduction in the torque variation between the maximum and minimum values.

A typical armature for a commercial dc starter motor is shown in figure 14-8, page 78. This armature has many loops of heavy wire with additional commutator

Fig. 14-8 Starter motor armature.

segments to reverse the current in individual loops at the proper time. The improvement in smoothing the torque, due to the additional loops, can be compared to the addition of cylinders in an automobile engine.

ACHIEVEMENT REVIEW

Select the *best* answer for the problems below to make each statement true. Place the letter of your answer in the space provided.

1. The Right-hand Motor Rule is usually used to determine _____
   a. flux density.   d. flux direction.
   b. direction of conductor movement.   e. induced current.
   c. conductor speed.

2. In the diagram shown, conductor movement will be _____
   a. upward.
   b. downward.
   c. to the right.
   d. to the left.
   e. constant.

   N
   ⊙
   S

3. Torque on a single loop of wire in a magnetic field is _____
   a. constant.
   b. strength.
   c. the same as field flux.
   d. the turning force.
   e. never at a maximum value.

4. The amount of torque on a current-carrying conductor in a _____
   magnetic field depends upon
   a. the amount of current in the conductor.
   b. the direction of the magnetic field between the two poles.
   c. the current direction in the conductor.
   d. the direction of rotation.
   e. the Left-hand Rule.

5. To obtain motor action, current is supplied to a loop of   _____
   wire in a magnetic field by
   a. slip rings.
   b. split rings.
   c. a commutator.
   d. brushes.
   e. brushes and a commutator.

6. If another loop of wire is added to make a double-loop   _____
   armature,
   a. the torque becomes steadier.
   b. the magnetic field decreases in value.
   c. loop current direction is affected.
   d. the torque becomes less smooth.
   e. commutator segments must be reduced in number.

7. The principle of motor action is   _____
   a. a conversion of mechanical energy to electrical energy.
   b. a conversion of chemical energy to electrical energy.
   c. a conversion of electrical energy to mechanical energy.
   d. an unpredictable phenomenon.
   e. predictable with the Left-hand Rule.

Summary Review of Units 11-14

15

OBJECTIVE

- To evaluate the knowledge and understanding acquired in the study of the previous four units.

For items 1–10, select the word or phrase at the right to make each incomplete statement true. Place the letter of the selected answer in the space provided.

1. Magnets are made of iron and iron _____
2. When like poles of a magnet are placed close to each other, they _____
3. The magnetic lines of force in the field of a magnet are referred to as _____
4. Lines of force are closed _____
5. The properties of a current-carrying conductor can be described as _____
6. Field strength is expressed by _____
7. The direction of the flux around a conductor carrying current is determined by the direction of the _____
8. The strength of an electromagnet depends on the amount of current, the number of turns, and the _____
9. The kind of magnetism that is present in a piece of iron that is brought near the north pole of a magnet is called _____
10. The twisting force created by a single-loop armature motor is referred to as _____

a. circles.
b. magnetic intensity.
c. current.
d. alloys.
e. core.
f. repel.
g. loops.
h. flux density.
i. voltage.
j. induced magnetism.
k. flux.
l. magnetic.
m. attract.
n. torque.
o. neutralize.
p. induced voltage.
q. wire size.
r. work.
s. energy.

In items 11–30, select the *best* answer to make each statement true. Place the letter of your answer in the space provided.

11. Strong magnetic fields may best be established in a core made of _____
    a. air.
    b. steel.
    c. nickel.
    d. cobalt.

12. Lines of force _____
    a. never cross.
    b. often cross.
    c. cross only under certain circumstances.
    d. are unpredictable.

13. Induced voltage in a conductor is a function of field _____
    strength, conductor length, and
    a. conductor cross-sectional area.
    b. conductor wire size.
    c. an external voltmeter.
    d. conductor speed.

14. An alternating emf can be obtained from a generator with _____
    a. a commutator.   c. slip rings.
    b. a split ring.   d. a load resistor.

15. Generally, the output voltage from a single-loop, _____
    two-pole generator is
    a. large.
    b. small.
    c. very steady.
    d. adequate for most applications.

16. The kind of magnetism that remains in a core material when _____
    the magnetizing force is removed is called
    a. residual.   c. polarization.
    b. magnetic.   d. north-south.

17. Maximum voltage is developed in a single-loop generator _____
    armature where the loop conductors, in relation to the
    magnetic field, move
    a. perpendicular.
    b. parallel.
    c. at a low speed.
    d. away from the pole faces.

18. The alternating emf generated in the single-loop armature of a _____
    two-pole generator reverses once every
    a. revolution.   c. quarter revolution.
    b. half revolution.   d. two revolutions.

19. The voltages that are induced in the armature conductors of _____
    a dc generator are
    a. unidirectional.
    b. direct.
    c. alternating.
    d. at a constant magnitude.

20. The direction of movement of the conductor, shown in the
    diagram, is   _____
    a. downward.
    b. upward. | S | ⊗ | N |
    c. to the right.
    d. to the left.

21. The strength of an electromagnet depends mainly on the   _____
    a. voltage and size of the wire.
    b. current and the size of the wire used.
    c. voltage and number of turns.
    d. current and number of turns.

22. The voltage generated in a single-loop generator armature is   _____
    a. dc.   c. alternating.
    b. pulsating dc.   d. unidirectional.

23. The commutator of a dc generator   _____
    a. reverses the direction of the current in the armature.
    b. changes ac to dc within the armature.
    c. keeps the current in one direction in the load circuit.
    d. acts only as a sliding electrical contact.

24. Brushes are required on a dc motor to   _____
    a. provide a sliding contact.
    b. change the direction of current in the armature.
    c. support the commutator.
    d. change the direction of the current in the external circuit.

25. An electromagnetic field is used in dc motors to   _____
    a. commutate the current more easily.
    b. reverse the rotation.
    c. give the motor greater speed.
    d. give the motor higher torque.

26. Any magnet may have   _____
    a. two kinds of poles.   c. three kinds of poles.
    b. many kinds of poles.   d. one kind of pole.

27. Lines of magnetic force   _____
    a. form closed loops pointing out at the N pole.
    b. point from N to S within the magnet.
    c. start at the S pole and end at the N pole outside the magnet.
    d. cross at the center of the magnet.

28. The armature of a commercial dc generator has many loops   _____
    of wire and many commutator segments to
    a. provide a high resistance path.
    b. balance the armature.
    c. cause the current in the loops to be steady.
    d. cause a high uniform output voltage.

29. In a dc generator, the direction of the emf induced in the   _____
    armature depends on the
    a. number of lines of force.
    b. speed of the armature.
    c. action of the commutator.
    d. magnetic polarity of the field poles.

30. The commutator of a dc motor   _____
    a. acts as a sliding contact only.
    b. reverses the current in the armature conductors.
    c. acts as bearing points for the commutator.
    d. reverses the current in the load circuit.

16 Typical Bell Circuits

OBJECTIVES

After studying this unit, the student will be able to

- construct typical low-voltage bell circuits.
- describe the signaling action of devices, such as bells, buzzers, pushbuttons, and bell transformers.

Practically every electrical installation includes some type of signaling circuit. This unit covers the procedures used to connect typical bell circuits. Information is also given on how various types of signaling devices operate, as well as information on pushbuttons, bell transformers, and wire used in low-voltage signaling circuits.

RULES FOR BELL CIRCUITS

It is easy to wire simple low-voltage bell circuits if three rules are followed.

1. Connect a conductor from one side of the voltage source to the bell.
2. Connect a conductor from the other side of the voltage source to the control point or pushbutton.
3. Connect a conductor from the pushbutton to the bell which is to be controlled.

Figure 16-1 shows a bell controlled from one pushbutton using a bell-ringing transformer as a source of power. Note that one connection is made directly from one terminal of the transformer output to the bell. A second conductor connects the other terminal of the transformer output to the switch or pushbutton. The third connection is from the pushbutton to the second bell terminal. The result is a simple series circuit.

BELL AND BUZZER CIRCUITS

A bell may be controlled from several different pushbuttons in different locations. In this case the same three wiring rules apply as shown in figure 16-2, page 86. Notice that the pushbuttons are in parallel, and the pushbutton combination is in series with the bell.

Several signaling devices, each controlled from a separate pushbutton, may be used on the same bell transformer. However, to distinguish among the signaling devices operated from different pushbuttons, each device must have a different tone.

INPUT
120 V ac

Fig. 16-1 Simple bell circuit.

Fig. 16-2 Bell controlled from several locations.

Most homes have pushbuttons at both the front and rear doors. In general, the bell or chime is controlled from the pushbutton located at the front door. The buzzer, or a bell with a different tone, is controlled from the pushbutton located at the rear door. Note that the circuit in figure 16-3 is a series-parallel circuit.

Combination units containing a bell and buzzer encased as one are commonly used. This type of signaling device has three terminals, figure 16-4. The terminals are connected as follows: the center terminal to the transformer, the right-hand terminal to the pushbutton at the front door, and the left-hand terminal to the pushbutton at the rear door.

Fig. 16-3 Bell-buzzer circuit.

Fig. 16-4 Connections for combination bell-buzzer unit.

PUSHBUTTONS

The device used to open and close a bell circuit is the pushbutton. It consists of a metal cover that holds a small insulated button in place on top of a spring contact. When the button is depressed, the circuit is closed; when the button is released, the circuit is open. In effect, this is a small, single-pole, normally open switch. Pushbuttons are usually mounted on the wooden doorframe at the front and rear door entrances. Although there are many types of surface and flush-mounted pushbuttons for different applications, they are all basically the same in operation.

THE DOORBELL

The interior connections of a typical doorbell are shown in figure 16-5. Two small coils of insulated wire mounted on iron cores form an electromagnet. Current passes from terminal A to terminal E by way of the electromagnets, the contact points B and C, and the armature labeled D. The electromagnets and contact points are ungrounded while the armature and terminal E are grounded to the case.

When this circuit path is energized, the two coils become electromagnets and attract the armature toward the iron cores. This, in turn, causes the hammer to strike the gong and, at the same instant, causes contacts B and C to separate by the action of the moving armature. The circuit is now open and the coils no longer attract the armature. The spring now returns the armature to its original position and the circuit is again closed. This process is repeated each time the hammer strikes the gong and continues as long as the bell circuit is energized. Because this cycle of operation occurs rapidly, the armature, contact spring, and hammer vibrate rapidly.

THE BUZZER

To distinguish between the tone of two signaling devices which are controlled from different pushbuttons, one bell and one buzzer can be used. As shown in figure 16-6, the buzzer does not have a gong and hammer but, otherwise, it is identical to the bell in connections and operation.

COMBINATION BELL AND BUZZER

Figure 16-7 is a diagram of a combination bell and buzzer mounted in one enclosure. The upper coil is the electromagnet for the bell, and the lower coil is the electromagnet for the buzzer.

DOOR CHIMES

Many residential installations use chimes rather than bells and buzzers. Instead of a harsh ringing or buzzing sound, a musical chime or tone is produced. Chimes are available in single-note, two-note, repeater-tone (where both notes continue to sound as long as the pushbutton is depressed), and the more elaborate eight-note (four-tube) styles. For the eight-note chime, contacts

Fig. 16-5 Vibrating bell.

Fig. 16-6 Buzzer.

Fig. 16-7 Combination bell and buzzer.

on a motor-driven cam are arranged to sound the notes of a simple melody in a pre-determined sequence.

The latter two styles are particularly useful in homes where three entrances are provided. For example, the chime can be connected so that the eight-note melody (or repeater tone) indicates the front door, the two notes indicate the side door, and the single note indicates the rear door.

BELL TRANSFORMER

The transformers required to operate door chimes are usually of greater capacity than the transformers used with bells and buzzers. The voltage output of a chime transformer is usually between 10 to 24 volts with a rating of from 5 to 20 volt-amperes. A bell transformer usually has a voltage output of 6 to 10 volts with a rating of from 5 to 20 volt-amperes.

Chime transformers for homes are available with a 16-volt rating. Transformers which give a combination of voltages such as 4, 8, 12, and 24 volts also can be obtained. The usual requirement is that the secondary current of this type of transformer must not exceed 8 amperes under short-circuit conditions. A further requirement is that the secondary voltage must not exceed 30 volts under open-circuit conditions.

BELL WIRE

The wire used for low-voltage bell and chime circuits is commonly called bell wire, annunciator wire, or thermostat wire. One type of wire consists of a copper conductor covered with two layers of cotton wrapped in opposite directions. These layers can be tied off to prevent unraveling of the insulation at terminals or splices. Both layers of this cotton wrapping are impregnated with paraffin. Another type of wire is insulated with a thermoplastic compound. Because of the low voltages involved, paraffin and thermoplastic insulations are satisfactory. Since the current required for bell circuits is small, No. 18 AWG conductors are usually used.

Multiconductor cables of two, three, or more single wires contained within a single protective overall covering are available. It is this type of cable that is commonly used in electrical installations because it gives a neat appearance to the wiring and because there is less danger of damage to individual wires. Color-coding of conductors within cables makes circuit identification easy.

Bell wire and cable may be fastened directly to surfaces with insulated staples, or may be installed in raceways. The particular requirements of the installation will determine how the conductors are to be attached.

NATIONAL ELECTRICAL CODE RULES

In general, bell wires with low-voltage insulation must not be installed in the same enclosure or raceway with lighting or power conductors. The outer jacket on nonmetallic-sheathed cable, UF cable, and armored cable meets this requirement. Bell wires must not come closer than two inches to open lighting or power conductors unless the bell wires are permanently separated from the lighting or power conductors by some approved type of insulation in addition to the insulation on the wire. Furthermore, bell wire with low-voltage insulation may not enter an outlet box or switch box containing lighting or power conductors unless a metal barrier is used to separate the two types of wiring. The National Electrical Code should be consulted for further information on specific installations.

ACHIEVEMENT REVIEW

1. In the diagram, the pushbutton in Plant A is to operate the buzzer in Plant B. The pushbutton in Plant B is to operate the bell in Plant A. Only one source of supply is available at Plant A and only three wires may be used between the two plants. Complete the wiring diagram.

In items 2–6, select the *best* answer to make the statement true, and place the letter of the answer in the space provided.

2. In a vibrating bell, the sound is made by the armature being pulled to the   _____
   a. pushbutton.   d. grounded case.
   b. contact points.   e. terminals.
   c. electromagnets.

3. The circuit that is formed by a single bell controlled from one location is called a   _____
   a. series circuit.   d. combination circuit.
   b. parallel circuit.   e. Norton circuit.
   c. series-parallel circuit.

4. The usual AWG number for bell wire is   _____
   a. 8.   b. 12.   c. 18.   d. 24.   e. 31.

5. Bell wires with low-voltage insulation may enter an outlet box containing power conductors if   _____
   a. there is a great voltage difference.
   b. the bell wire is of normal size.
   c. a chime circuit is being wired.
   d. a metal partition is used.
   e. your foreman thinks it's all right.

6. Bell transformers
   a. can be used with only one signaling device.
   b. can be used with several signaling devices.
   c. should be placed at the front and rear doors.
   d. should have different tones.
   e. can only be used with a single pushbutton.

17 Switch Control of Lighting Circuits

OBJECTIVES

After studying this unit, the student will be able to

- describe the various types of switches used for the control of lighting circuits.
- list the ratings and categories of switches.
- discuss switch circuits and describe the use of various types of switches.

The electrician installs and connects various types of lighting switches. Therefore, it is necessary to know how each type of switch operates and the standard connections for each type of switch. The electrician must understand the meaning of the current and voltage ratings marked on lighting switches and be familiar with the National Electrical Code requirements for the installation of these switches.

TOGGLE SWITCH

The most frequently used switch in lighting circuits is the toggle flush switch or snap switch, figure 17-1. When mounted in a flush switch box, the switch is concealed in the wall with only the insulated handle or toggle protruding.

Four types of toggle switches are available: single pole, three way, four way, and double pole. A three-way, flush-type toggle switch is shown in figure 17-2.

Fig. 17-1 Toggle flush switch. *Courtesy Pass and Seymour, Inc.*

Fig. 17-2 Three-way, flush-type toggle switch. *Courtesy Pass and Seymour, Inc.*

Ratings of Switches

Underwriters' Laboratories classifies toggle switches used for lighting circuits as *general-use snap switches* and divides these switches into two categories.

- *Category 1* — ac/dc general-use snap switches may control resistive loads not to exceed the ampere rating of the switch at rated voltage; may control inductive loads not to exceed one-half the ampere rating of the switch at rated voltage; and may control tungsten filament lamp loads not to exceed the ampere rating of the switch at 125 volts when marked with the letter T. (This latter condition is imposed because a tungsten filament lamp takes a very high current the instant the circuit is closed and, thus, subjects the switch to a severe current surge.)

    The ac/dc general-use snap switch usually is not marked ac/dc. However, it is always marked with the current and voltage rating, such as 10A-125V, or 5A-250V-T.

- *Category 2* — ac general-use snap switches are to be used on alternating-current circuits only. They may control resistive, inductive, and tungsten filament lamp loads not to exceed the ampere rating of the switch at 120 volts; and may control motor loads not to exceed 80 percent of the ampere rating of the switch at rated voltage, but not exceeding two horsepower. Category 2 switches are marked ac in addition to current and voltage ratings, such as: 15A, 120–277V ac. These switches also can be marked ac only. The 277-volt rating is required on 277/480-volt systems.

Single-pole Switch

A single-pole switch is used when a light or group of lights, or other load, must be controlled from one switching point. This type of switch is connected in series with the ungrounded or hot wire feeding the load. Figure 17-3 shows a typical application of a single-pole switch controlling a light from one switching point. Note that the 120-volt source feeds current directly through the switch.

Fig. 17-3 Circuit with single-pole switch: feed at switch.

In figure 17-4, the 120-volt source feeds current directly to the light outlet. This results in a two-wire cable with black and white wires being used as a switch loop between the light outlet and the single-pole switch. The National Electrical Code permits the use of a white wire in a single-pole switch loop. However, the black conductor must connect between the switch and the load. Note that this requirement is satisfied in figure 17-4.

Fig. 17-4 Circuit with single-pole switch: feed at light.

Figure 17-5, page 94, shows another application of a single-pole switch control. The feed is at the switch which controls the light outlet. The convenience outlet is independent of the switch.

Double-pole Switch

A double-pole switch is used when it is necessary to break (open) both conductors of a circuit. This circuit for a lamp on a gasoline dispensing island is illustrated in figure 17-6, page 94.

Three-way Switches

A three-way switch has one terminal, called the common terminal, to which the switch blade is always connected. In addition, there are two other terminals. These are called the traveler wire terminals. In one position, the switch blade is connected between the common terminal and one of the traveler terminals. In the alternate position, the switch blade is connected between the common terminal and the other traveler terminal. Figure 17-7, page 95, shows the two positions of the three-way switch. Note that the three-way switch is actually a single-pole, double-throw switch.

The three-way switch has no ON or OFF position. As a result, there are no ON or OFF markings on the switch handle. The three-way switch can be identified further

Fig. 17-5   Ceiling outlet controlled by single-pole switch with live convenience receptacle: feed at switch.

Fig. 17-6 Application of a double-pole switch.

Fig. 17-7 Two positions of a three-way switch.

by its three terminals. The common terminal is darker in color than the two traveler wire terminals which are natural brass in color.

The three-way switch is used when a light or group of lights, or other load, must be connected from two different switching points. To accomplish this, two three-way switches are used, as shown in figure 17-8.

Fig. 17-8 Circuit with three-way switch control.

In figure 17-8, note that one light is to be controlled from either of two switching points. The feed in this circuit is at the first switch control point. Often, it is convenient to be able to control a hall light from either an upstairs or downstairs location, or a garage light from either the house or the garage.

Figure 17-9 shows a different circuit arrangement using three-way switch control with the feed at the light. It is necessary to use the white wire in the cable as part of the three-way switch loop in this circuit. The black wire is used as the return wire to the light outlet.

Fig. 17-9 Circuit with three-way switch control: feed at light.

Figure 17-10 represents another arrangement for three-way switch control. The feed is at the light with cable runs from the ceiling outlet to each of the three-way switch control points which are located on each side of the light outlet.

Fig. 17-10 Circuit with three-way switch control: feed at light.

Four-way Switch Control

A four-way switch can be compared with a double-pole, double-throw switch. It is similar to a three-way switch in that it has two positions and neither of these positions is ON or OFF. As a result, the four-way switch has no ON or OFF markings on the switch handle. Two positions of a four-way switch are shown in figure 17-11.

Fig. 17-11 Two positions of a four-way switch.

The four-way switch is used when a light or group of lights, or other load, must be controlled from more than two switching points. The switch connected to the source and the switch connected to the load must be three-way switches. At all other control points, four-way switches are used.

Figure 17-12 illustrates a typical circuit where a lamp is controlled from any one of three switching points. Care must be used in connecting the traveler wires to the proper terminals of the four-way switch. Always make sure that the two traveler wires from one three-way switch are connected to the two terminals on one side of the four-way switch while the two traveler wires from the other three-way switch connect to the two terminals on the other side of the four-way switch.

Fig. 17-12 Circuit with switch control at three different locations.

ACHIEVEMENT REVIEW

1. What is the most commonly used style of lighting switch?

2. List four types of lighting switches.

   a. _____   c. _____

   b. _____   d. _____

3. To control a group of lights from one control point, what is the most practical type of switch to use?

   _____

4. What type of switch is used if it is necessary to control a group of lights from two different control points?

   _____

5. Complete the connections in the following arrangement so that both ceiling light outlets are controlled from the one single-pole switch. Assume the installation is in cable.

6. Complete the connections in the following arrangement so that the ceiling outlet may be controlled from either three-way switch.

7. Show the connections for a ceiling outlet which is to be controlled from any one of three switch locations. The 120-volt feed is at the light.

8. Determine what type and rated switch is required to control five 300-watt tungsten filament lamps on a 120-volt circuit.

   _____

   _____

   _____

   _____

In problems 9–11, select the *best* answer to complete the statement, and place the letter of the answer in the space provided.

9. If a light is to be controlled from three different locations, the   _____
   following switches must be used:
   a. two three-way and one four-way.   d. two double-pole.
   b. two four-way and one three-way.   e. three single-pole.
   c. three three-way.

10. A three-way switch is similar to a   _____
    a. four-way switch.   d. single-pole, double-throw switch.
    b. double-pole, double-throw switch.   e. double-pole, single-throw switch.
    c. single-pole, single-throw switch.

11. When a tungsten filament lamp is turned on, the current   _____
    through the switch is initially
    a. low, but builds up.   d. dependent on the switch
    b. very high.   resistance.
    c. moderate.   e. steady.

18 Wiring Materials

OBJECTIVES

After studying this unit, the student will be able to

- list the various types of wiring materials available.

- explain the advantages and limitations of each wiring material.

  Six standard materials are used in wiring:

- nonmetallic-sheathed cable

- armored cable

- flexible steel conduit

- rigid conduit

- thinwall conduit or electrical metallic tubing

- surface metal raceway

These various types of wiring materials are discussed in this unit, including their advantages, limitations, and applications. The National Electrical Code should be referred to when studying this unit. The articles of the National Electrical Code which are related to the sections of this unit must be thoroughly understood so that the electrician has a complete understanding of why these wiring materials may or may not be used in different applications and locations.

NONMETALLIC-SHEATHED CABLE

Nonmetallic-sheathed cable is available with two or three current-carrying conductors in sizes ranging from No. 14 through No. 2 with copper conductors, and in sizes No. 12 through No. 2 with aluminum conductors. Color-coding of these conductors is black and white for two-wire cable, and black, white, and red for three-wire cable. This cable is also available with a ground wire which is usually an uninsulated copper conductor. This conductor is to be used for grounding ONLY. Insulation on the current-carrying conductors may be Type T or TW.

Underwriters' Laboratories, Inc., lists nonmetallic-sheathed cable in two classifications:

1. Type NM cable may be used for both exposed and concealed work in normally dry locations. It has an overall flame-retardant and moisture-resistant covering. It may be fished (drawn through) in the hollow spaces of masonry block or tile

walls where such walls are not exposed to excessive dampness. Masonry that is in direct contact with the earth is considered a wet location.

Type NM cable shall not be installed where exposed to corrosive fumes or vapors and shall not be embedded in masonry, concrete, fill, or plaster.

2. Type NMC cable may be used for both exposed and concealed work in dry, moist, damp, or corrosive locations. It has an overall flame-retardant, moisture-resistant, fungus-resistant, and corrosion-resistant covering. It may be embedded in plaster or run in hollow spaces of masonry walls. Type NMC cable is commonly installed in buildings where a highly corrosive atmosphere is present.

The National Electrical Code lists various locations where Types NM and NMC cable shall NOT be used. These locations include, for example, service entrances, places of public assembly and hazardous areas, among others. The Code should be consulted for the complete list and/or exceptions.

Both types of nonmetallic-sheathed cable shall be strapped or stapled not more than 12 inches from a box or fitting and at intervals not exceeding 4 1/2 feet; shall be protected against physical damage where necessary; shall not be bent to a radius less than five times the diameter of the cable; and are for use on circuits of 600 volts or less.

Nonmetallic cable will be supplied without any ground wire unless WITH GROUND WIRE is specified. Nonmetallic cable has various trade names, such as Braidx, Cresflex, Loomwire, and Romex. Figure 18-1 shows the two-wire cable with and without a ground wire.

Special connectors, such as the one shown in figure 18-2, are used to secure nonmetallic cable to outlets such as fuse boxes and device boxes, figure 18-3.

Fig. 18-1 Nonmetallic cable with and without a ground wire.

Fig. 18-2 Nonmetallic cable connectors.

Some types of connectors are first securely fastened to the cable. The threaded section of the connector is then slipped through the knockout hole in the outlet box. Finally, the locknut is securely fastened to the connector on the inside of the outlet box.

Fig. 18-3 Device box for armored cable and nonmetallic-sheathed cable. *Courtesy National Electric Products Corporation*

When nonmetallic cable is used, it is necessary to remove the outer covering to make necessary connections in outlet boxes and switch boxes. This is done by slitting the braid with a knife as far back as necessary. The braid and paper removed from the wires are then cut off. In removing this outer braided covering, extreme care must be used so that the wires are not damaged.

Nonmetallic-sheathed cable is the least expensive wiring method to use. This cable is relatively light in weight and is easy to install. For these reasons, it is widely used for residential installations.

ARMORED CABLE (BX)

Armored cable, figure 18-4, is available with two, three, or four conductors that come in sizes from No. 14 AWG to No. 1 AWG, inclusive. Color coding is as follows: for two-wire cable, black-white; for three-wire cable, black-white-red; for four-wire cable, black-white-red-blue. Type AC cable is armored cable with conductors insulated with Code-grade rubber covered with a flame-retardant and moisture-resistant finish.

ANTISHORT BUSHING   CODE INSULATION

GALVANIZED STEEL ARMOR   BONDING STRIP   PAPER WRAPPER   COTTON BRAID   TINNED COPPER CONDUCTOR

Fig. 18-4 BX armored cable.

Since the development of thermoplastic insulations, armored cable is usually manufactured with Type T or Type TW insulation. This is called Type ACT cable.

Armored cable is required to have an internal bonding strip of either copper or aluminum in close contact with the armor for its entire length. This metal strip, plus the flexible steel armor, makes this cable desirable when a grounded system is required. The armor also adds mechanical protection to the conductors.

Whenever a connection is made in an outlet box or a switch box, it is necessary to cut the metal armor back a distance of 6 to 8 inches from the end of the cable. To prevent any damage to the conductors, a fiber bushing must be inserted between the steel armor and the wires at the point where the armor is cut. Figure 18-5 shows the steps required to remove the armor and insert the fiber bushing (called an antishort bushing) between the conductors and the armor. Connectors for BX armored cable are shown in figure 18-6.

Cut Armor and Slide Off   Insert Bushing Between   Untwist Conductors and Tear
                                 Paper Wrap and Armor   Off Paper Close to Bushing

Fig. 18-5   Removing armor and inserting bushing. *Courtesy National Electric Products Corporation*

Fig. 18-6   Connectors for BX armored cable. *Courtesy National Electric Products Corporation*

In general, Type AC and Type ACT cable may be used on circuits up to 600 volts; may be used for open and concealed work in dry locations; may be fished through walls and partitions, and may be embedded in the plaster finish on masonry walls or run through the hollow spaces of such walls if these locations are not considered damp or wet. This cable shall be secured within 12 inches from every outlet box or fitting and at intervals not exceeding 4 1/2 feet, and shall not be bent to a radius of less than 5 times the diameter of the cable.

BX cable is not approved for use underground, and it cannot be embedded in masonry, concrete, or the fill of buildings during construction. This cable cannot be installed in any location which is exposed to weather, oil, gasoline, or other materials which have a deteriorating effect on rubber insulation.

Fig. 18-7 Type ACL or BXL flexible armored cable.

The National Electrical Code recognizes another type of armored cable which contains conductors that are insulated with varnished cambric. This type of cable is listed as Type ACV and is generally installed in commercial and industrial buildings.

Another standard BX armored cable, called Type ACL or BXL flexible armored cable, has a protective lead sheath covering the conductors instead of a paper covering, figure 18-7. After the lead sheath is applied, an interlocked steel strip is wrapped in a spiral over the entire assembly. This cable can be obtained in wire sizes from No. 14 AWG to No. 4 AWG in either two- or three-wire cable. The conductors are color coded for easy identification. In the two-wire cable the colors are black and white; in the three-wire cable, they are black, red, and white. One type of device box for armored cable is shown in figure 18-8.

Type ACL (BXL) cable is used for circuits of 600 volts or less. It may be run in concrete and masonry, and it may be used in locations that are exposed to weather, continuous moisture, or excessive humidity. This type of cable also may be used in places where it is exposed to oil, gasoline, or other materials which have a deteriorating effect on rubber or other types of insulation.

Fig. 18-8 Device box for armored cable, nonmetallic-sheathed cable, and loom.

FLEXIBLE METAL CONDUIT

Flexible metal conduit, figure 18-9, is sometimes called Greenfield cable. This conduit is similar to BX armored cable. It is formed with a single strip of galvanized metal, wound in a spiral on itself, and interlocked so as to provide maximum strength with greatest flexibility. The electrician must pull wires through this conduit. This flexible conduit is measured by its inside diameter and is listed in sizes from 3/8 inch to 4 inches, inclusive. The National Electrical Code should be referred to for the rules covering the use of flexible metal conduit.

Fig. 18-9 Flexible metal conduit. *Courtesy National Electric Products Corporation*

Whenever a rigid raceway system requires a flexible section to meet difficult installation conditions, flexible metal conduit may be used. This conduit is used to provide a flexible raceway to adjustable equipment such as a motor mounted on an adjustable base for a belt drive. It is recommended for temporary wiring installations where local codes specify that wiring be in metallic conduit. It is approved for most locations except wet locations, hoistways, storage battery rooms, hazardous locations, or where conditions may have a deteriorating effect on the conductor insulation.

A wiring material very similar to flexible metal conduit is liquidtight flexible metal conduit. This conduit has an outer liquidtight jacket over the armor. This jacket makes the conduit suitable for locations subject to oil, water, certain chemicals, and corrosive atmospheres. The conduit is available in sizes from 3/8 inch to 4 inches.

RIGID CONDUIT

Rigid conduit is an extremely permanent type of material, and the wires must be pulled through as with flexible metal conduit. This conduit pipe is annealed and heat treated so as to permit easy bending. Conduit comes in ten-foot lengths and may be obtained with either a galvanized or a black-enameled finish.

Galvanized Conduit

Galvanized conduit has a heavy, but smooth and uniform, coating of zinc applied to both the exterior and interior surfaces. After the zinc is applied, a coating of insulating lacquer is baked on all interior and exterior surfaces to produce a smooth raceway through which wires may be pulled with a minimum of effort. The combination of the heavy zinc coating and the lacquer coating protects the conduit from moisture and corrosive fumes. Galvanized conduit is also available without the lacquer coating.

Black-enameled Conduit

Black-enameled conduit is made from the same high-grade metal used for galvanized conduit. The black enamel is applied to the conduit and is then baked and oxidized to a hard, close-adhering finish which protects the pipe against corrosion. The black enamel also makes the interior of the pipe smooth and wires may be pulled through with a minimum of effort.

Fig. 18-10 Galvanized rigid conduit and coupling. *Courtesy National Electric Products Corporation*

Fig. 18-11 Black-enameled conduit and coupling. *Courtesy National Electric Products Corporation*

Fig. 18-12 Elbows for rigid conduit. *Courtesy National Electric Products Corporation*

Each ten-foot length of conduit, whether galvanized or black enameled, is threaded on both ends. One coupling is furnished with each 10-foot length of conduit. Figure 18-10 shows galvanized conduit and figure 18-11 shows black-enameled conduit.

Factory-bent elbows are available for all sizes of rigid conduit from 1/2 inch to 6 inches, figure 18-12. However, electricians generally do their own bending at the job site. Most of the small conduit sizes are bent with hand benders and hickeys (bending devices), while the larger sizes are bent with the aid of hydraulic benders.

Plastic-coated conduit is available. It is resistant to the severe corrosive atmospheres that are found in certain areas of sewage treatment plants, metal refineries, tanneries, and similar locations.

Aluminum conduit is available and has several advantages over other types of conduit. It is

- only about one-third the weight of galvanized conduit.

- corrosion resistant.

- nonmagnetic, resulting in less voltage drop per given length as compared to metal conduit. Therefore, the power loss is reduced.

A complete line of aluminum elbows, straps, locknuts, bushings, condulets, and other fittings is available from various manufacturers. These fittings are manufactured to conform to the requirements of the National Electrical Code.

More detailed information on particular fittings can be obtained from manufacturers' catalogs and specifications.

Conduit is manufactured in sizes ranging from 1/2 inch to 6 inches. The rigid conduit size is always determined by the internal diameter and even this value is slightly larger than the rated size. The following table gives the internal and external diameters in inches for each trade size of rigid conduit.

Conduit Fittings

Rigid conduit is secured to junction boxes, outlet boxes, and fuse boxes by locknuts and end bushings.   Figure 18-13 illustrates various types of boxes for conduit.

Size in Inches	Diameter in Inches		Threads per Inch
	Internal	External	
1/2	0.622	0.840	14
3/4	0.824	1.050	14
1	1.049	1.315	11 1/2
1 1/4	1.380	1.660	11 1/2
1 1/2	1.610	1.900	11 1/2
2	2.067	2.375	11 1/2
2 1/2	2.469	2.875	8
3	3.068	3.500	8
3 1/2	3.548	4.000	8
4	4.026	4.500	8
5	5.047	5.563	8
6	6.065	6.625	8

(A) 4-INCH OCTAGON BOX

(B) 4-INCH SQUARE BOX, 1 1/2 INCHES DEEP

(C) DEVICE BOX

Fig. 18-13 Various types of boxes for conduit. *Courtesy National Electric Products Corporation*

Fig. 18-14 Locknut. *Courtesy National Electric Products Corporation*

Fig. 18-15 Conduit end bushing. *Courtesy National Electric Products Corporation*

Figure 18-14 illustrates a locknut used with rigid conduit. This locknut is turned on the threaded end of the conduit pipe with the teeth formed by the notches facing toward the box. The conduit is then slipped through the knockout hole and a metal end bushing, figure 18-15, is screwed to the end of the conduit as tightly as possible. The locknut is then tightened solidly against the outside wall of the outlet box. The teeth of the locknut must bite into the metal of the outlet box to assure that the conduit pipe is securely bonded to ground.

The end bushing on the end of the conduit thread secures the conduit to the inner wall of the outlet box and protects the wires from possible damage from the edge of the conduit. When using end bushings made entirely of insulating material (such as plastic), a locknut shall be installed both inside and outside the enclosure.

Condulet Fittings

A complete line of condulet fittings is available for any installation problem involving rigid conduit. Condulet fittings, figure 18-16, page 110, are threaded and can be tightened securely to the threaded end of the conduit. Condulet fittings are available in the same sizes given for conduit pipe. For example, 1/2-inch pipe requires 1/2-inch condulet fittings.

Condulet fittings have wiring chambers large enough to permit splicing and taping. Condulets with wire-hole covers may be used as outlets for motors and control equipment. Certain types of condulet fittings permit the mounting of flush wiring devices, while other types are used for light outlets. Condulet fittings are specified by the trade size and a letter identification.

Rigid conduit provides maximum protection to conductors. This type of raceway also acts as an effective ground for equipment. Rigid conduit is a standard wiring method and may be used in nearly all situations. It is used for concealed wiring in buildings where the wiring is buried in concrete or masonry. When wiring is to be exposed and subject to mechanical damage, rigid conduit is a satisfactory wiring method.

Rigid metallic conduit is one of the few raceways permitted by the National Electrical Code for use on systems operating at more than 600 volts. With the proper

TYPE A

TYPE LB

TYPE B

TYPE LR

TYPE C

TYPE E

TYPE T

TYPE X

Fig. 18-16 Condulet fittings. *Courtesy Crouse-Hinds Company*

fittings, rigid conduit may be installed in hazardous locations. It may NOT be installed in or under cinder fill unless it is protected on all sides by a layer of noncinder concrete at least 2 inches thick, or is protected by corrosion-resistant material suitable for the purpose. (Certain types of cinders form sulfuric acid when in the presence of moisture. This acid is very corrosive on steel conduit.)

Bends in conduits containing conductors without lead sheathing shall have a radius not less than *six* times the diameter of the conduit. Bends in conduits containing conductors with lead sheathing shall have a radius not less than *ten* times the trade diameter of the conduit.

The National Electrical Code rules cover all types and installations involving rigid conduit. Specific tables of the Code refer to the number of conductors which can be run in various sizes of rigid conduit, electrical metallic tubing, and flexible metallic conduit.

ELECTRICAL METALLIC TUBING

Electrical metallic tubing is a thinwall rigid metallic conduit. The walls of this conduit are substantially lighter in weight than the walls of rigid conduit pipe. Metallic tubing, therefore, does not offer the same protection against mechanical damage, or the corrosive action of water or chemicals as does rigid metal conduit.

Thinwall metallic tubing is *not* threaded, figure 18-17. Compression couplings are used for joining lengths of metallic tubing, and compression connectors are used to secure metallic tubing to outlet and junction boxes.

Fig. 18-17 Electrical metallic tubing. *Courtesy National Electric Products Corporation*

Electrical metallic tubing comes in 10-foot lengths and is made in sizes ranging from 1/2 inch to 4 inches. The internal diameter is slightly larger than the stated size. The table lists a few of the available sizes in inches of electrical metallic tubing and the internal and external diameters for each size.

Metallic tubing is light and easy to handle. Furthermore, with the use of compression couplings and connectors, figures 18-18 and 18-19, this type of raceway can be assembled quickly. For most wiring jobs, time is important and minutes saved mean dollars earned.

Electrical metallic tubing is abbreviated EMT, and fittings used with this type of raceway are called EMT fittings. Figure 18-18 is a threadless coupling used to join lengths of electrical metallic tubing. Figure 18-19 shows the EMT fitting used as a connector to secure metallic tubing to outlet and junction boxes.

Electrical metallic tubing may be used for open or concealed work where it will not be subject to severe mechanical damage or to corrosive vapors. This tubing may not be used in cinder concrete or fill unless protected on all sides

Size in Inches	Diameter in Inches		Wall Thickness in Inches
	Internal	External	
1/2	0.622	0.706	0.042
3/4	0.824	0.922	0.049
1	1.049	1.163	0.057
1 1/4	1.380	1.510	0.065
1 1/2	1.610	1.740	0.065
2	2.067	2.197	0.065

Fig. 18-18   Coupling for electrical metallic tubing. *Photo Courtesy National Electric Products Corporation*

Fig. 18-19 Connector for electrical metallic tubing. *Photo Courtesy National Electric Products Division*

by a layer of noncinder concrete at least 2 inches thick, or unless the conduit is at least 18 inches under the fill.

Tubing smaller than the 1/2-inch size may not be used except under special conditions specified in the National Electrical Code. The maximum tubing size which may be used with any number or combination of conductors is the 4-inch size. Electrical metallic tubing cannot be used in interior wiring systems if the voltage is greater than 600 volts. The number of conductors permitted in EMT is the same as that of rigid conduit.

SURFACE METAL RACEWAYS

The surface metal raceway is a two-piece, flat metal raceway which can be mounted on ceilings and walls. The base or channel is securely fastened to the ceiling or wall surface by screws, toggle bolts, or rawl drives. The cover or capping is secured directly to the channel or base.

This type of raceway generally is used in office buildings, public buildings, and some industrial plants for making additions to existing installations or where future changes are probable. The surface metal raceway is neat in appearance and does not detract from the appearance of a room. The raceways are relatively small in size and can be used with special fittings to go around beams and corners.

The National Electrical Code permits surface metal raceways in dry locations for exposed or surface work. The raceway can be extended through dry walls, dry partitions, and dry floors if one continuous length of raceway is used throughout the concealed section.

The raceway cannot be used for concealed work in any location where it is subject to severe mechanical damage, or where the voltage is 300 volts or more between conductors unless the wall thickness of the raceway is not less than 0.040 inch. This type of raceway cannot be used where it may be subjected to corrosive vapors, or in hoistways and any hazardous location.

Two types of surface metal raceways are known as National Metal Molding, figures 18-20 and 18-21,

Fig. 18-20 National metal molding. *Courtesy National Electric Products Corporation*

and Wiremold, figures 18-22 and 18-23. A complete line of fittings is available for each of these makes of raceway. Wiremold is available in standard lengths of 10 feet.

Fig. 18-21 National metal molding clip. *Courtesy National Electric Products Corporation*

Fig. 18-22 Wiremold. *Courtesy The Wiremold Company*

Fig. 18-23 Wiremold fittings. *Courtesy The Wiremold Company*

ACHIEVEMENT REVIEW

1. List five standard materials used in wiring.

    a. _____

    b. _____

    c. _____

    d. _____

    e. _____

2. State five places where nonmetallic-sheathed cable cannot be used.

a. _____

b. _____

c. _____

d. _____

e. _____

3. What is the difference between types AC and ACL armored cable? _____

_____

_____

_____

_____

4. List five locations where BX cable cannot be used.

a. _____

b. _____

c. _____

d. _____

e. _____

5. State the difference between flexible armored cable and flexible metal conduit.

_____

_____

_____

_____

6. List two applications of flexible metal conduit.

a. _____

_____

b. _____

_____

7. A piece of armored cable measures 3/4 inch in diameter. Determine the minimum radius to which this cable may be bent. _____

_____

8. Explain why rigid galvanized conduit can be used in practically any wiring application. _____

_____

_____

9. What is the principal use of surface metal raceway? _____

_____

_____

_____

_____

10. State two advantages of rigid aluminum conduit as compared to rigid galvanized conduit of the same size.

a. _____

b. _____

In items 11 through 16, select the *best* answer to complete each statement, and place the letter of your answer in the space provided.

11. The advantage of electrical metallic tubing over rigid conduit   _____
(same size) is that it
   a. is stronger.
   b. is lighter.
   c. has a larger external diameter.
   d. resists mechanical damage better.
   e. resists the corrosive action of water better.

12. Condulet fittings are used for   _____
   a. flexible armored cable.
   b. surface metal raceway.
   c. electrical metallic tubing.
   d. nonmetallic cable.
   e. rigid conduit.

13. An antishort bushing is used on   _____
   a. armored cable.
   b. surface metal raceway.
   c. electrical metallic tubing.
   d. nonmetallic cable.
   e. rigid conduit.

14. The maximum distance from a box that nonmetallic-sheathed   _____
cable shall be stapled will not exceed
   a. 1/2 ft.   d. 5 times the diameter.
   b. 1 ft.   e. 4 1/2 ft.
   c. 2 ft.

15. For a given set of electrical installation requirements, the   _____
least expensive material to use is
   a. surface metal raceway.
   b. thinwall conduit.
   c. nonmetallic-sheathed cable.
   d. armored cable.
   e. flexible metal conduit.

16. The size of rigid conduit is determined by the   _____
   a. inside diameter.
   b. outside diameter.
   c. the radius of bend that can be made.
   d. style of pipe.
   e. type of coating.

19 Remote Control Systems for Lighting Circuits

OBJECTIVES

After studying this unit, the student will be able to

- explain the principles of basic remote control systems.

- list and describe the devices used in remote control systems.

A remote control wiring system uses controlling devices such as relays. A dial telephone system is probably one of the best known remote control wiring systems. In the telephone system, relays at a distant point are operated by turning the dial or pushing the buttons on the telephone.

Remote control systems have been developed to control lighting circuits, appliances, and other equipment in various situations. These remote control systems consist of low-voltage relays (24 volts) which operate 120-volt contacts from low-voltage controlling switches.

The low-voltage remote control system makes it possible to have multiswitch control with only a small increase in cost. Switches are easily installed and the installation of one switch or several usually does not present any problem. The same type of low-voltage switch is used whether light outlets are controlled from one, two, or more locations.

Special cables are not required as in three-way and four-way switch connections. Low-voltage remote control circuits may be installed in two- or three-wire cable using No. 18 AWG wire which is low in cost and easy to install.

SWITCH CONTROL

The switch used in the low-voltage remote control system is a single-pole, double-throw momentary contact switch which is normally open. This type of control switch is approximately one-third the size of a standard single-pole switch. It may have three terminals, four terminals, or three color-coded lead wires. Regardless of the type of switch used, all will function as a single-pole, double-throw momentary contact switch. Figure 19-1 illustrates one type of low-voltage switch.

Fig. 19-1 Low-voltage switch.

Connections are made to this low-voltage switch as follows: terminal No. 4 is connected to the 24-volt transformer source; terminal No. 1 is not connected to the switch contacts, but is used only for connection purposes so that it is not necessary to have any splices; and the two other terminals connect to the relay or to other low-voltage switches.

LOW-VOLTAGE RELAY

A split-coil relay is used in low-voltage remote control systems to operate contacts in the 120-volt lighting circuit, figure 19-2. One coil closes the 120-volt circuit and the other coil opens the contactors in the 120-volt circuit. This relay will operate on 24 volts. It is equipped with two No. 12 leads for the 120-volt circuit and three colored leads for the 24-volt connections. The relay, figure 19-3, is small enough to be mounted from the inside of a standard outlet box through any 1/2-inch knockout hole opening. This leaves the two high-voltage leads inside the outlet box while the low-voltage end is outside the box. The wall of the outlet box serves as a partition between the high and low voltages. The two high-voltage leads inside the outlet box are connected like a single-pole switch.

Fig. 19-2 Relay connections.

Fig. 19-3 Relay. *Courtesy The General Electric Company*

TRANSFORMER

Only one transformer is required to operate this low-voltage remote control system. The transformer steps down the lighting circuit voltage of 120 volts to 24 volts for the low-voltage control circuit. The rating of the transformer used is not any greater than 100 volt-amperes. When overloaded, the output voltage of the transformer decreases, resulting in less current. Thus, the transformer has current-limiting characteristics to counteract any overload.

CONDUCTORS

No. 18 AWG solid conductors are used for remote control systems. These conductors usually are made into either two-conductor cable or three-conductor cable. Insulation on the individual conductors may be the double-cotton covering or it may be a plastic material. Regardless of the type of insulation used, the individual conductors of either the two-conductor or the three-conductor cable can be identified easily.

INSTALLATION PROCEDURE

In new construction, the 120-volt system is roughed-in as usual, except that all switch legs or switch loops are omitted. Although each manufacturer may number or color code his devices in a different manner, the electrician need only remember that three low-voltage conductors are necessary between the relay and the switch, and that two low-voltage conductors carrying the low-voltage supply from the transformer must be run to either the relay or to the switch. These wires are spliced like color to like color, or certain colors may be assigned to correspond to the numbers on the relays and switches.

TYPICAL WIRING INSTALLATIONS

The wiring diagrams (figure 19-4 and figures 19-6 and 19-7 on pages 120 and 121) show several of the most common installations. Many other combinations are possible to meet particular requirements.

Figure 19-4 represents a low-voltage remote control system where one 120-volt light is controlled from one switch point. A transformer is shown in figure 19-5.

Another common application is shown in the wiring diagram in figure 19-6, page 120, in which each light has individual switch control. That is, switch 1 controls light 1 and switch 2 controls light 2.

Fig. 19-4 One light controlled from one switch point.

Fig. 19-5 Transformer. *Courtesy The General Electric Company*

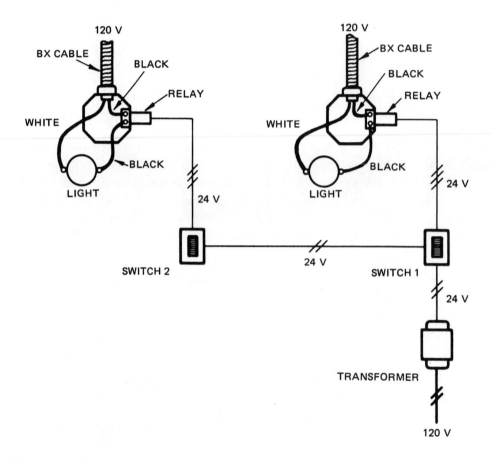

Fig. 19-6 Two lights with individual switch control.

Figure 19-7 illustrates a wiring diagram for two lights which are both to be controlled from any one of three switch points.

For specific remote control wiring requirements, the National Electrical Code should be consulted for rules and regulations that govern the installation. The low-voltage remote control system is not subject to the same Code restrictions as the standard 120-volt system.

Fig. 19-7 Two lights controlled from any one of three switch locations.

ACHIEVEMENT REVIEW

1. State two benefits of remote control lighting systems.

   a._____

   b._____

2. What is the function of each of the two low-voltage coils in the relay?_____

   _____

   _____

3. What type of switch is used as a control in a low-voltage system?_____

   _____

   _____

In items 4–9, select the *best* answer to make the statement true, and place the letter of your answer in the space provided.

4. The type of switch used in remote control systems is   _____
   a. a momentary contact switch.
   b. quite large.
   c. a double-pole switch.
   d. a single-throw switch.
   e. a double terminal switch.

5. To control three lights connected in parallel, from two different   _____
   points,
   a. three relays are required.
   b. only one transformer is required.
   c. two transformers are required.
   d. only one switch is necessary.
   e. one relay and one switch are necessary.

6. Code restrictions for low-voltage remote control systems   _____
   a. are the same as those for the standard 120-volt system.
   b. specify that No. 12 AWG wire be used in the low-voltage
      part of the circuit.
   c. do not exist.
   d. are different from those for the standard 120-volt system.

7. If a short circuit occurs in a light socket which is part of a   _____
   remote control system, and a transformer with current-
   limiting characteristics is being used, the
   a. current through the short will have a high value.
   b. output voltage from the transformer will increase.
   c. current in the short will be very low.
   d. current and voltage will remain the same as they were
      before the short occurred.

8. The cable used in the low-voltage part of remote control systems   _____
   a. is the same as cable used in standard 120-volt systems.
   b. requires special insulation.
   c. must allow fuses or circuit breakers to be easily installed.
   d. is low in cost as compared to cable used in standard 120-volt
      systems.

9. For remote control systems,   _____
   a. the switch used has a built-in relay.
   b. the transformer is sometimes built into the relay.
   c. the relay always must have a separate outlet box.
   d. multiswitch control is usually not possible.

10. Draw a remote control wiring diagram to control one light from two switch points with the devices shown below.

120 V

L

"ON" COIL

"OFF" COIL

ON

OFF

SWITCH 2

ON

OFF

SWITCH 1

24 V   120 V

TRANSFORMER

20 Summary Review of Units 16-19

OBJECTIVE

- To evaluate the knowledge and understanding acquired in the study of the previous four units.

In items 1–13, fill in the word(s) that will make the statement correct.

1. The device that is used to close and open a bell circuit is the _____.

2. The most practical type of switch to use to control a group of lights from one location is the _____.

3. The type of switch that is used to control a group of lights from two locations is the _____.

4. The type of switch which must be used together with three-way switches to control a group of lights from three or more locations, is the _____.

5. When a bell is to be operated by several pushbuttons, the pushbuttons must be wired, with respect to one another, in _____.

6. Nonmetallic-sheathed cable must be supported by staples within a certain distance from the box. That distance is _____.

7. Armored cable must be supported by straps or staples at intervals not to exceed _____.

8. Referring to wiring materials, the letters EMT mean _____.

9. The antishort bushing is used for the type of cable called _____.

10. The type of switch that is used to open both conductors of a circuit at the same time is a _____.

11. The type of conduit that is sometimes called Greenfield cable is _____.

12. The name for the type of fitting that is used for rigid conduit is the _____.

13. The inside diameter of a conduit is used to specify the trade _____.

In items 14–20, select the *best* answer to make the statement true, and place the letter of the answer in the space provided.

14. An end bushing is used for _____
    a. flexible steel conduit.
    b. armored cable.
    c. nonmetallic-sheathed cable.
    d. condulets.
    e. rigid conduit.

15. Door chimes
    a. are not used much anymore.
    b. require greater capacity transformers than those used
       for bells and buzzers.
    c. are only available as two-note devices.
    d. require transformers with less capacity as compared to
       transformers used with bells and buzzers.
    e. require low-voltage relays with 120-V contacts.

16. One area where armored cable *cannot* be used is
    a. underground.
    b. embedded in the plaster finish on masonry walls.
    c. in a concealed dry location.
    d. inside a building, but in the open in a dry location.
    e. through walls and partitions.

17. Surface metal raceway
    a. cannot be used around corners.
    b. is neat in appearance.
    c. is relatively large in size.
    d. is used in hazardous locations.
    e. is used in concealed locations.

18. The type of switch that is used in remote control systems is a
    single-pole, double-throw switch, and is
    a. normally closed.
    b. a two-terminal switch.
    c. a momentary contact switch.
    d. energized by a relay.
    e. sometimes used in 120-volt circuits.

19. Electrical metallic tubing, sometimes called thinwall conduit,
    a. utilizes compression couplings.
    b. may be used in any application where rigid galvanized conduit
       is used.
    c. can be bent with as small a radius as desired.
    d. has threads cut at the ends to secure fittings.
    e. is heavy as compared to black-enameled conduit.

20. In remote control circuits,
    a. fuses are required in the low-voltage portion.
    b. lights may be controlled from any number of switch locations.
    c. the transformer may be of any volt-ampere rating.
    d. the transformer should be designed so that when overloaded
       the output voltage increases.
    e. the number of transformers required is directly proportional to
       the number of lights to be controlled.

21. With the devices shown below, draw a wiring diagram showing the proper connections for a single-family dwelling.

22. In the following diagram, the pushbutton in Plant A is to operate the buzzer in Plant B. The pushbutton in Plant B is to operate the bell in Plant A. Only one source of supply is available and only three wires may be used between the two plants. Complete this wiring diagram.

23. Show the connections for a ceiling outlet which is to be controlled from either of two switch locations. The 120-volt feed is at the light. Be sure to color code all wires.

S3

S3

BX CABLE

LAMP

SOURCE

24. Using the devices shown below, light L1 is to be controlled from one control
point, and light L2 is to be controlled from two control points. One transformer
is to be used. Draw the wiring diagram that will accomplish the necessary control.

Acknowledgments

The staff at Delmar Publishers:

    Sponsoring Editor: William W. Sprague
    Senior Editor: Marjorie A. Bruce
    Project Editor: Frances Larson

    Reviewer: Dennis Gable

Title Page Photographs: Apprentice Committee
                           Joint Industry Board of the Electrical Industry
                           Flushing, New York

The author and editorial staff at Delmar Publishers are interested in continually improving the quality of this instructional material. The reader is invited to submit constructive criticism and questions. Responses will be reviewed jointly by the author and sponsoring editor. Send comments to:

Delmar Publishers Inc.
Sponsoring Editor
Technical Division
50 Wolf Road
Albany, New York 12205

Index